AF560680

TRANSLOCATION
IN
PLANTS

TRANSLOCATION IN PLANTS

By

Dr. Shubhrata R. Mishra
Department of Botany
Vikram University
Ujjain (M.P)

DISCOVERY PUBLISHING HOUSE
NEW DELHI-110002

First Published-2004

ISBN 81-7141-890-2

Published by

DISCOVERY PUBLISHING HOUSE
4831/24, Ansari Road, Prahlad Street,
Darya Ganj, New Delhi-110002 (India)
Phone: 23279245 • Fax: 91-11-23253475
E-mail:dphtemp@indiatimes.com

Printed at:

ARORA OFFSET PRESS
Laxmi Nagar, Delhi-92.

PREFACE

"Translocation in Plants" has been carefully compiled and edited to meet the long felt needs of increasingly large number of those who have to deal with the different aspects of the transport of various substances from one part of plant to the other. It provides a balanced and integrated treatment of the entire field transport system. The title is intelligible to the educated layman but it deals with some complex ideas. It is an adequate text for all requirements in this area for most university students. Special efforts have been made to explain ideas in non mathematical terms. The primary aim throughout has been clarity, simplicity and the high standard. It will definitely prove to be an authoritative work to teachers, students and research workers in the field of transport system.

Though, the author has taken special care to present a current account, yet he is fully aware of his limitations, and the readers may come across the shortcomings of various types. For all types of shortcomings he extends his due apology.

The author has freely consulted other standard books and research papers while preparing the manuscript, so the author claims no originality of the work.

The author expresses his thanks to his friends and colleagues whose continuous inspirations have invited him to bring out this text.

The author expresses his gratitute to Mr. Wasan and staff of M/s Discovery Publishing House for their whole hearted co-operation in the publication of this book.

Constructive criticisms and suggestions for improvement of the book will be thankfully acknowledged.

Author

CONTENTS

5. Phloem Translocation **259—296**

1

STEM

TYPES OF STEMS

Stems are commonly *siphonostelic protosteles* occurring in living plants only among the fevus and some other pteridophytes and usually possess secondary growth. Stems differ greately in the amount and arrangement of primary vascular tissues and in the amount of secondary tissues. The primary vascular tissue ranges in amount from that of a solid cylinder of considerable thickness to that of a few small strands forming isolated bundles. The various conditions apparently represent stages in evolutionary progress where there has been a thinning of the cylinder of primary vascular tissue, a breaking up of the cylinder into longitudinal strands,and changes in the arrangement of the resulting bundles so that these no longer form a cylindrical series.

Whatever the amount and the arrangement of the primary xylem, the secondary vascular tissues may form a solid cylinder enclosing it and it this way develop an unbroken vascular cylinder even from a series of unevenly placed strands. The amount and the arrangement of secondary xylem also vary from that of the complete cylinder of indefinite thickness formed in typical perennial woody axes to be isolated thin strands of certain types of annual herbaceous stems and to the condition in other plants where secondary growth is absent. Secondary vascular tissues, like primary, have been reduced in evolutionary modification, the cylinder being first thinned radially and then broken up tangentially. In forms most specialized in this respect, no secondary vascular tissues are formed. All stages in these changes in primary and secondary vascular tissues are represented among living plants. Great variety is found in stem structure.

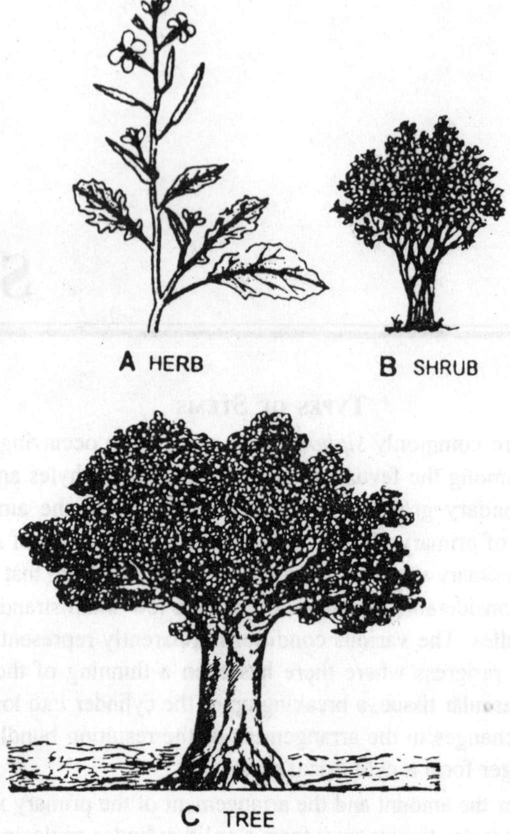

Fig. 1.1. Plant habit: A—herb, B—shrub, C—tree.

Woody Stem

Perennial woody plants present apparently simple stem structure. In these an unbroken layer of secondary vascular tissue sheaths a more or less continuous cylinder of primary xylem. Variations in the structure of this cylinder, range from cylinders that are unbroken except by leaf and branch gaps to those consisting of discrete bundles often complex in arrangement. The simplicity of the secondary cylinder, concealing the basic primary structure, seems to make the entire structure simple. Even where the primary cylinder is broken only by leaf and branch gaps prominent "bundles" stand out as protoxylem-containing ridges in the otherwise very thin cylinder. These ridges represent the downward continuation of the leaf traces and other early maturing parts of the cylinder. In other plants the cylinder consists of more or less closely placed bundles of varying size–with proportionate amounts of protoxylem—without connecting primary xylem. The areas between these primary strands are ray-like projections of the pith on the inside and pericycle on the outside. They are soon closed by the cambium which builds secondary vascular tissue in these spaces. Woody, perennial stems, such as those of some palms and other large monocotyledons, may be developed without secondary growth.

Herbaceous Stem

Herbaceous plants have no distinctive anatomical structure. Annual stems are commonly supposed to have their vascular tissue—whether it is primary only, or primary plus secondary—characteristically in the form of discrete bundles arranged in a cylinder, but this condition is not typical of herbaceous plants. Among dicotyledons, the majority of herbaceous forms possess cylinders of vascular tissue that are complete, except for the presence of leaf and branch gaps. Such conditions obtain not only in coarse and stout stems, such as those of many composites, mints, and legumes, but even in slender delicate stems, such as those of species of *Veronica* and *Stellaria*. The herbaceous stem with discrete bundles

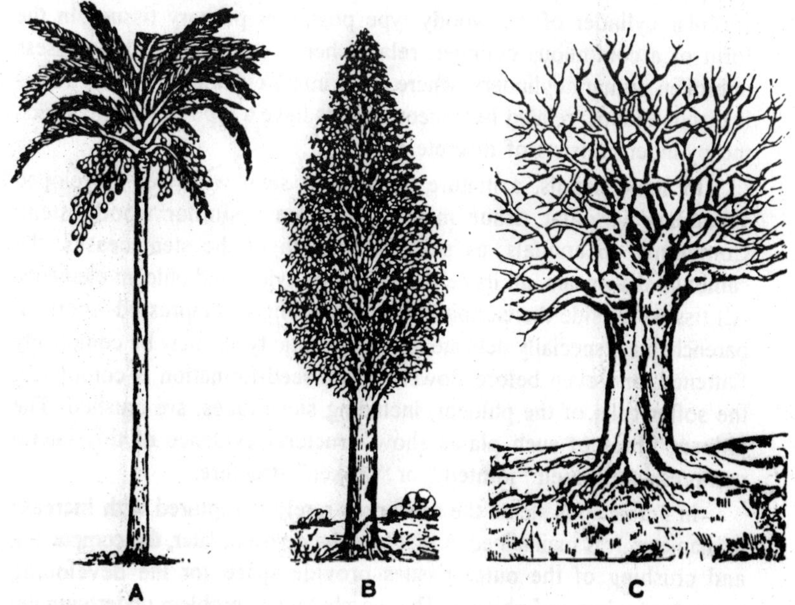

Fig. 1.2. Tree types: A—caudex, B—excurrent.

is not to be considered typical of annual stems. It represents, clearly, the extreme stage of reduction of vascular tissues. Although infrequent in occurrance, it is found in various families and in both stout and slender plants.

In some forms, for example, in species of *Trifolium*, *Geum*, and *Agrimonia*, the lower part of the stem may have a complete cylinder of vascular tissue, and the upper part separate bundles. The condition in monocotyledons is essentially the same as that in extreme dicotyledonous herbs; here the bundles may lie in a ring, but more frequently they are distributed through the stele according to a plan determined by the number of leaf traces, lead arrangement, and other factors.

Herbaceous stems, except those of the extreme type where the cambium fails to unite the primary skeleton, are structurally like woody stems; differences are those of period of persistence, not of basic structure. The herbaceous stem is one in which cambial growth is limited to one season or part of one season, or is lacking. The amount of vascular tissue, primary or secondary, is not necessarily less in a mature herbaceous stem than in the one-year-old stem of a closely related woody plant of the same type. The type of stem in a given herb depends upon the type of stele present in the woody ancestral forms. Where the vascular cylinder of the woody type possesses primary tissues in the form of a continuous cylinder, related herbaceous forms also possess unbroken primary cylinders; where the primary cylinder of woody forms is discontinuous, related herbaceous plants have a type of stele in which the cylinder consists of discrete bundles.

In some details, a mature herbaceous stem with well developed secondary vascular tissue may differ from a similar woody stem. Commonly, in annuals, as seasonal growth of the stem ceases, the cambium disappears, all its cells naturing as xylem and phloem elements. All tissues outside the cambium become strongly compressed—cortical parenchyma, especially delicate, photosynthetic types, may be completely flattened; and even before flowering and seed-formation is completed, the softer cells of the phloem, including sieve tubes, are crushed. The phloem fibres of such plants show structural evidence of this lateral compression in their "jointed," or "broken" structure.

In herbaceous stems the epidermis rarely is ruptured with increase in diameter. It is maintained at first by slow division; later, the compacting and crushing of the outer tissues provide space for the developing secondary xylem and phloem. The morphological problem underlying the evolutionary changes in structure from the typical woody stem to the extreme herbaceous stem lies outside the scope of this book. From the evidence of comparative morphology and of the fossil record, it is clear that in angiosperms the herbadeous type of stem has been derived from the woody, undoubtedly independently in many families. In a few families, such as the *Berberidaceae*, woody types have been derived from herbs. Woody plants of this type have many features in common with herbs, especially structure of the primary skeleton and histological structure of the xylem.

Monocotyledonous Stem

Secondary growth of the usual type is typically lacking throughout the body of monocotyledons but vestiges of cambial activity in the bundles both of the stem and the leaves have been found in nearly all groups. The stele is broken up into bundles which are commonly distributed throughout the axis; the endodermis is lacking, and the limits of cortex, pericycle, and pith are often indistinguishable, since bundles are scattered throughout. In many forms, as in most grasses, there is present a central region—which may or may not represent the pith morphologically—in which no bundles occur; in other forms a more or less readily separable cortical region may also be seen. The vascular system of most monocotyledons is highly complex.

The leaf-trace bundles are numerous and follow various types of courses in their descent, uniting in different ways with other strands. A common condition is that where all bundles are common bundles. The traces, upon entering the stem from the leaf, penetrate deeply, the median traces more deeply than the lateral, and then in their descent return toward the periphery. The downward course may be vertical, or the traces may swing laterally and become oriented in various ways. Each common bundle sooner or later fuses with other similar bundles. The anastomoses occur chiefly at the nodes,and in some groups, such as the grasses, are abundant and largely restricted to that region.

Secondary Growth in Monocotyledons

The type of secondary tissue formed in the monocotyledons is very different from that formed in other groups. The cambium does not produce phloem on the outside the xylem toward the inside in the normal way but forms on the inner side amphivasal or collateral bundles in parenchymatous ground tissue commonly called *conjunctive tissue.* The bundles are usually without definite arrangement but may lie to some extent in radial rows. Occasional anastomoses occur. The extracambial tissues formed are small in amount and parenchymatous in natrue. In the cambium layer proper, the divisions are largely tangential, and the

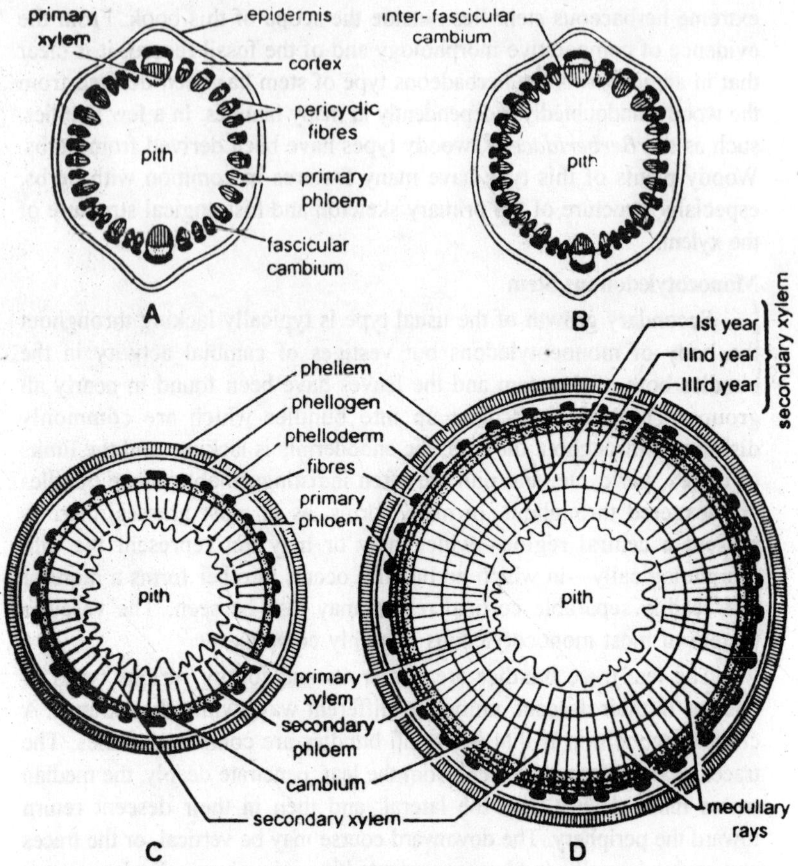

Fig. 1.3. Secondary growth in dicotyledonous stem. A—primary tissues only. B—Formation of inter-fascicular cambium. C—Secondary xylem and phloem are formed. D—Cross section of a three year old stem.

cells so formed are consequently radially arranged. This order is usually evident in the conjunctive tissue and aids in the separation of the secondary from the primary tissues in which the interfascicular parenchyma has no definite arrangement. It is not evident in the tissues of the bundles because of the method of development of these strands.

A very small group of cambium derivatives divide longitudinally—at first anticlinally and periclinally, then haphazardly —forming a strand of cells that mature as xylem and phloem. The cells that form the tracheids become 15 to 40 times their original length; the other cells elongate little or not at all. The tracheids may be scalariform, a type rare in secondary

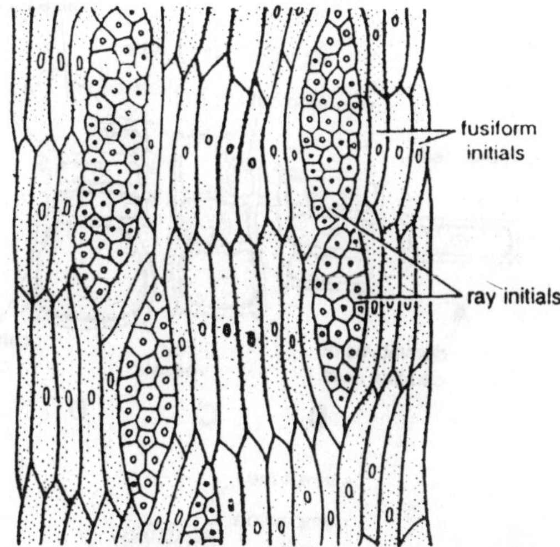

Fig. 1.4. Vascular cambium showing fusiform initials and ray initials.

tissue. The mature bundles differ from the primary bundle in the small amount of phloem and the absence of annular and spiral protoxylem cells.

In some genera differences in relative abundance of bundles and in wall thickness of the conjuctive cells set off weakly limited growth rings. The relation of these rings to annual growth is unknown. Secondary tissue of this type is indefinite in amount, as is that of the normal type, but usually develops slowly, and large trunks are not commonly formed. Increase in diameter of this type occurs, in the arborescent Lilliflorae, for example, *Dracaena*, *Yucca*, *Aloe*, *Cordyline*; some of the palms; rarely in herbs, as in *Veratrum*; and in fleshy parts of some of the *Dioscoreae*. The thickening which takes place in the bases of some palm stems is not due to the activity of a definite cambium layer but is, rather the result of gradual increase in size of cells and of intercellular spaces, and rarely, of the proliferation of strands of tissue to form new fibres; it represents long continuing primary growth.

Vine Type of Stem

The stem of vines are of both structural types. Many vines, such as *Vitis*, *Celastrus*, and *Sclanum Dulcamara*, have vascular steles in the form of woody cylinders; others, such as Clematis, *Humulus*, and *Pisum*, have vascular tissues arranged in the form of a ring of bundles. These bundles are separated by rays of parenchyma, which in many forms are increased, like the bundles, by cambial growth. In such forms, as in the type of herbaceous stem with discrete bundles, the xylem and phloem

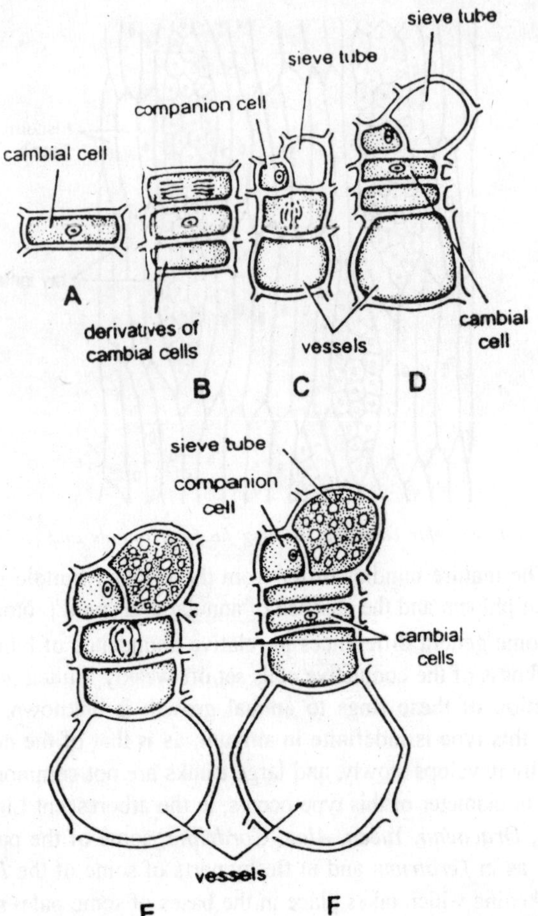

Fig. 1.5. Stages in the differentiation of secondary xylem and secondary phloem elements from a cambial cell.

are highly specialized in structure—vascular rays are commonly absent; the vessels are porous and of large diameter and great length; tracheids and fibres are proportionately few; sieve tubes are of the highest type; and phloem parenchyma and fibres are scarce or lacking.

Many vines show not merely these tissue specializations, but also possess anomalous general structure. In this group fall *Aristolochia* and *Menispermum*, which, unfortunately, are often cited as examples of typical stem structure, and used to demonstrate the origin of a woody from an herbaceous stem. In the stem of the young herbaceous vine, such as that of *Pisum*, *Apios*, and *Adlumia*, the vascular bundles are separated by wide segments of parenchyma. In these stems the cambium may be restricted to the bundles, as in Adlumia, or may from a complete cylinder by extension across the interfascicular rays of parenchyma. This interfascicular cambium may be vestigial, as in the upper parts of the pea vine, forming no secondary tissue, or a very few vascular cells, as at the base of the pea vine; or it may be very active, increasing the parenchymatous rays at the same rate as the fascicular cambium builds up the bundles, as in *Clematis*.

In such plants as the last, the type of structure found in the one-year stem is maintained as the stem becomes woody and perennial. Yet, since the bundles increase in tengantial extent as secondary growth continues, the stem apparently becomes a woody stem. The bundles are, however, still separated, though by wedges of secondary rather than primary parenchyma. These radial plates or rays of tissue commonly extend unbroken from node to node, often for several internodes and constitute an important feature of vine structure. Such rays, either of primary structure, or of both primary and secondary structure, are prominent structural features of many vines, both annual and perennial. Similar parenchymatous segments are present also in some vines with complete woody cylinders, and constitute one of the prominent modifications of such steles found in lianas. Since they consist of soft tissue, they are sometimes crushed as the stem becomes older, perhaps by the "play" of the bundles upon one another during the periods of lateral stresses to which a vine stem is peculiarly subject. Apparently, this type of structure represents one type of modification related to the mechanical requirements of vines.

Epidermis

The essential character of an epidermis is that it is a protective layer, or in other words, that it is a barrier between the internal organization cf the plant and the environment. This protection is not exclusively concerned with water loss, though that is a very important factor, and the surface layer of some submerged aquatics may be truly classed as an epidermis even though it is not a protection against desiccation. But the emphasis on the barrier function excludes from this category secretory surfaces such as those of glands, and absorptive surfaces such as those of the young root. Furthermore, a true epidermis is a primary layer, developed from the apical meristem, and the name does not apply to the exodermis of roots or the periderm of older stems, which are both secondary. On the other hand, certain specialized cells of the surface layer, such as hair cells and sclereids, which may not be primarily protective in function, are none the less morphologically part of the epidermis and we shall treat them as such.

Epidermal cells are usually described as tabular in form which means somewhat flattened radially, but when viewed from the outer surface they are seen to be nearly always vertically elongased, and they are by no means always radially flattened. In outline they vary greatly, but the variation is much less in the epidermis of the stem that in the leaf. Growth of the primary stem is chiefly elongation, and the epidermal cells in the young state are consequently elongate and narrow. The radial walls may be somewhat wavy, which is apparently the result of continued expansion of the cells after growth of the stem has ceased. The same factor may result in the outer surfaces being convex, domed or even papillate.

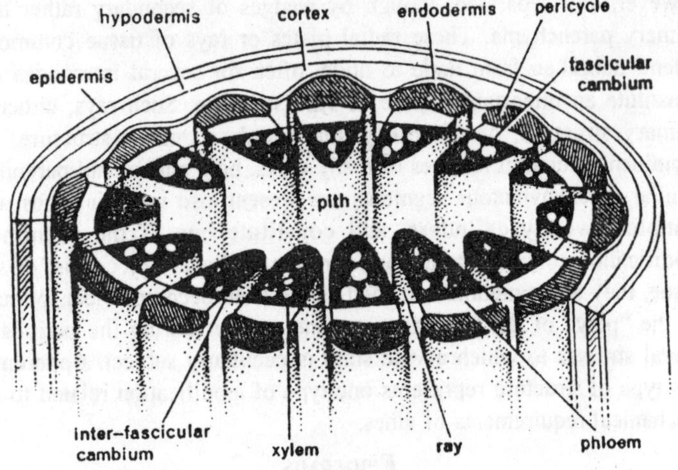

Fig. 1.6. Section of stem showing primary tissues.

The radial walls are not only undulating, when viewed from the exterior, but are thin and have numerous large pits. This implies that lateral movement of water in all directions in the epidermis must be relatively easy, and it is probable that a dangerous loss of water at any one point of the surface may be compensated by the tangential flow of water towards the threatened spot from other parts of the epidermis, which thus acts as a first line of defence, before the radial flow from the xylem can become effective. That epidermal cells do, in fact, part with considerable amounts of water at times, is shown by the great changes of volume they can withstand without injury, especially in leaves. When this occurs the radial walls shrink and the cells become flattened, but they rapidly expand when equilibrium is restored.

The epidermal cells of some plants (e.g., Erica and Daphne) have greatly thickened, mucilaginous inner walls, thus providing for increased water storage. The physiological as well as the mechanical cohension of the epidermal cells is increased by the absence of any spaces at the angles between them. The outer surface of the epidermis is covered by the cuticle. This layer, which varies in thickness in general relationship to the moisture conditions in the environment, is extra-cellular and continuous and therefore helps to bind the epidermal cells together. It is tough and elastic, only slightly permeable to either water or gases, and its protective value is very great. Its development begins very early, and it may sometimes be traced even on the tunica in the meristem.

Priestley has explained it as a non-volatile residue from the surface evaporation of sap, and while this would serve to account for its greater thickness in plants of dry habitats, it is difficult to accept so simple an explanation for cuticles on enclosed surfaces, such as inner wall of the ovary or on leaves within the winter bud. The cuticle contains no cellulose and is composed chiefly of the insoluble anhydrides of an unknown number of fatty acids, of which two, stearocutic and oleocutic acids, have been described. They probably reach the epidermis in the form of glycerides, i.e., as true fats, and are there decomposed, with the liberation of the free acids. The chief difference between cuticle and suberin, apart from their different location in the plant, seems to be the absence from the former of the phellonic acid which is an important constituent of the latter. Their reactions to microchemical test reagents are very similar. Continued growth of the epidermis, besides causing the wavy outlines of the cells, also throws the cuticle covering into wrinkles, which often form a minute pattern over the outer surface of each cell.

The cuticle often extends downwards for some distance along the radial wall, forming distinct wedges between the cells, but it never completely separates them. Below the cuticle proper lies the *cutinized layer*, which is formed by the cellulose outer walls of the cells, more or less impregnated with cutin. It is sharply separated from the true cellulose wall next to the cell lumer. The cutinized layer is not invariably present, but when it occurs it is frequently much thicker than the true cuticle and may be the chief protective layer. It often accompanies the cuticle in forming prominent wedges between the radial walls of the cells. Beneath the epidermis there is often, in plants of dry or exposed habitats, a second, or even a third, specialized layer, the walls of which are not cutinized but may be greatly thickened with cellulose or lignin, or else may be quite unthickened. It may augment both the protective and the water storage functions of the epidermis, and it is referred to either

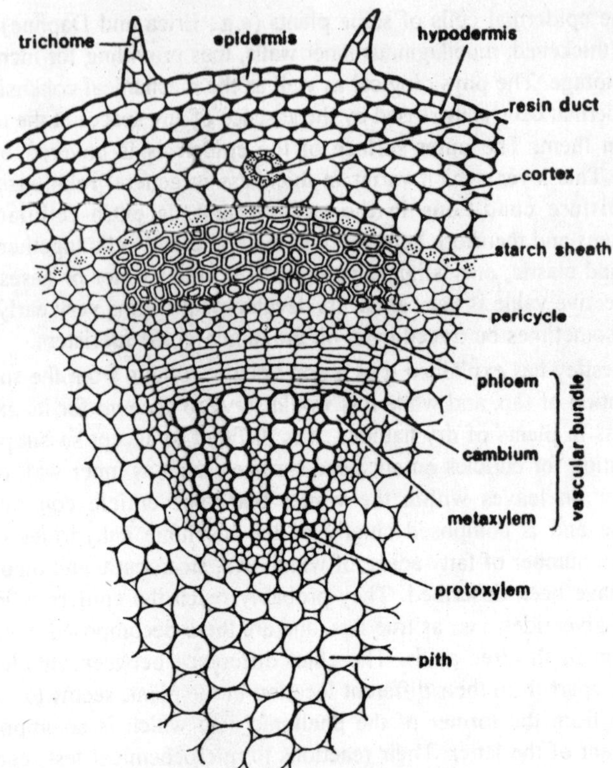

Fig. 1.7. T.S. stem of Helianthus.

separately as the *hypodermis*, or both layers together may be termed a *multiple epidermis*. Strictly speaking, the latter name should not be used unless it is known that both layers of cells have originated from the dermatogen by tangential cell divisions. As this cannot often be proved it is best to retain the general term hypodermis. The epidermis of young, green stems contains a number of *stomata*, which resemble those on the leaves. They are not normally so numerous as on the leaves, but in plants with reduced or abortive leaves they may be the only means for gas exchanges with the atmosphere, and in such cases they have the same importance in photosynthesis as those on normal leaves.

Epidermal Outgrowths

Outgrowths of the epidermis take many forms and are included under the general term *trichomes*, which covers not only true hairs, but other modified structures such as glands and prickles. They are, of course, not confined to the stem, but no distinction can be drawn between organs

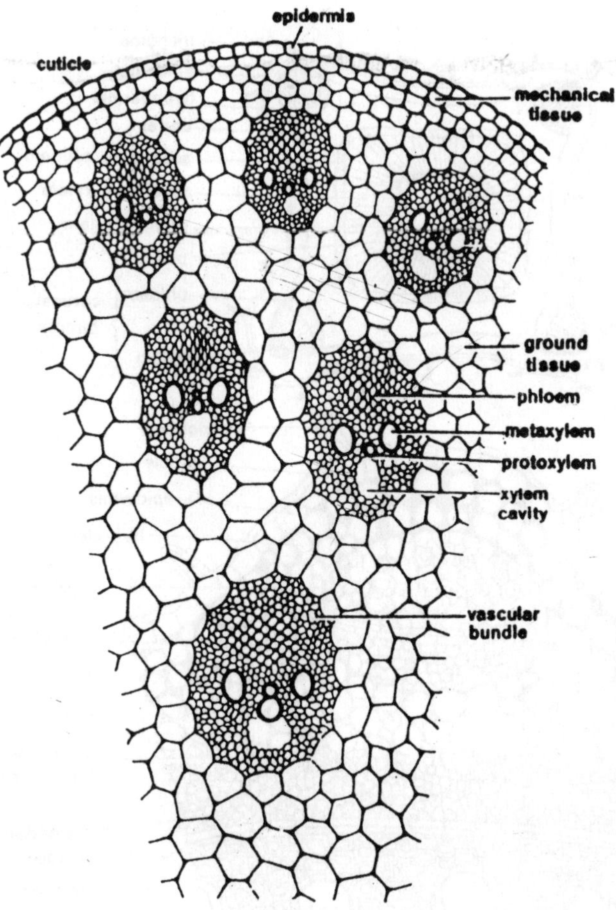

Fig. 1.8. T.S. of stem of Zea mays.

in this respect, since identical trichomes may be produced at all points on the surface of the shoot. Their distribution is, however, often restricted to definite lines or areas and may be a mark of distinction between related species. Despite the immense variety among trichomes, a complete anatomical series may be traced between the simplest, which are merely prolongations of a single epidermal cell, and massive structures which arise from groups of cells and involve also the sub-epidermal tissues and may even receive one or more vascular bundles.

These extreme cases have sometimes been distinguished as *energences*, but in spite of their apparent differences no sharp line can

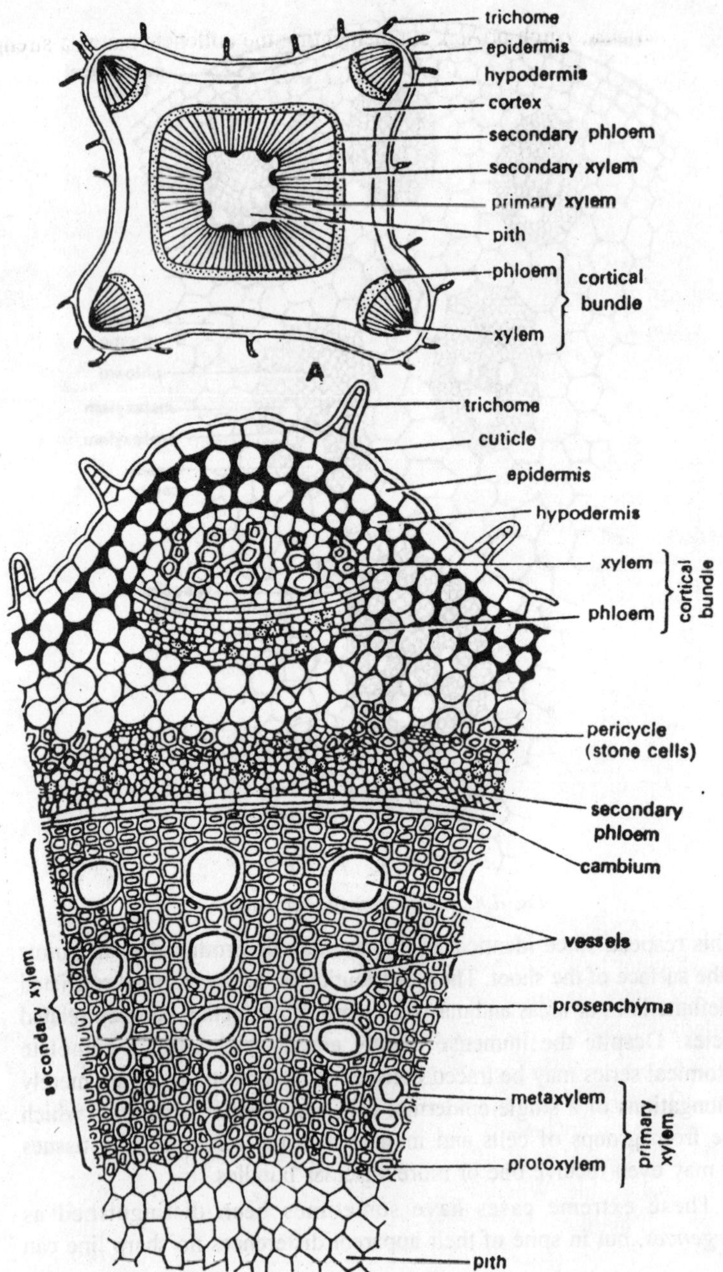

Fig. 1.9. T.S. of stem of Nyctanthes. A—Complete section. B—A part of section.

be drawn between them and the simpler trichomes. Hairs can be classified under three main types: simple, branched, and peltate. Simple and branched hairs may consist of one or of many cells. The branched types are sometimes quite elaborate, like small bushes; or they may be flattened, the branches forming a rosette around a short stalk-cell. The latter type is related to the peltate or scale hairs, which are similar to flattened branched hairs but with the branches cohering into a disc. These discs, if they overlap, as in the Eleagnaceae, from a complete protective armour over the surface of the epidermis. Secretory hair glands may be simply swollen cells which are usually water stores; or they may be compound, with long or short stalks bearing a head consisting of either one large secretory cell, or a more or less peltate group of cells.

Prickles are classified as trichomes, no matter how massive they may be, whenever it is clear that they arise from the epidermis and are not modifications of any other organ. As a matter of fact, the large prickles of *Rosa* and the vascular prickles on the fruit of Horse Chestnut are connected with simple hairs by all gradations of finer prickles. Prickles of all grades may even be present together on the same stem, as in the Burnet Rose (Rosa spinossissima), and it is notable that in some Roses the prickles at the nodes are often much larger and stronger than the others. The so-called "endogenous prickles" on the stems of some Palms are, however, really modified adventitious roots and are therefore properly classified as thorns. The very large and prominent spines of the Cacti are of doubtful nature, having been interpreted both as modified leaves and as trichomes. They grow little humps, called areolae, from which the lateral buds, if any, usually arise, and the balance of evidence points to the spines being really trichomes.

CORTEX

Between the epidermis and the stele lies the zone of the parenchymatous *cortex*. In stems it is rarely as broad as in the root, for the vascular tissues, probably for mechanical reason lie much nearer to the periphery than in roots. The underground stems of Monocotyledons usually have a fairly broad cortex, but monocotyledonous aerial stems, such as that of the Maize so often used as an anatomical type, are as a rule only temporary flowering shoots and may have a very narrow cortex or none at all. Such stems may be regarded as being either entirely stelar or else as having no stele. There is at any rate, in many cases, no stelar boundary visible. Where a cortex is present it often contains numerous and small marginal leaf traces. The outer zones of the cortex frequently form a *collenchyma* either in a continuous band or, in angular stems such as those of Labiatae and Umbelliferae, developed only below the ridges. Collenchyma is definitely a mechanical, strengthening tissue. It has this great advantage over sclerenchyma, that its cells are living and growing and hence it is able to adapt itself to the growth of young organs. It is chiefly characteristic of such organs, nevertheless in short-lived structures such as petioles or the herbaceous stems of Dicotyledons it may remain as the permanent mechanical tissue.

The aerial stems of Monocotyledons, on the other hand, depend for mechanical support more on the development of the sclerotic sheaths around their vascular bundles than on collenchyma. As collenchyma cells remain alive it follows that their cell walls cannot be uniformly thickened, for this would cut them off from all external supplies. The thickening material, which is of cellulose, is accordingly distributed in such a way that thin walls are left at certain parts of the cell to allow of intercommunication. Frequently the thickening appears only at the angles of the cells, while in other cases it may be all on one side or on two sides, an arrangement which also has the advantage of increasing the plastic and yielding character of the tissue as a whole.

The cells are somewhat elongated vertically, but never to the extent of woody fibres, and they may be regarded as mainly parenchymatous. Cortical cells usually contain functional chloroplasts since light can penetrate a distance up to 100 μ from the surface. True palisade tissue, comparable with that in the leaf, may be found in the cortex of some xerophytes, either in succulents of the Cactus type or in switch plants, like the Broom in fact, wherever the stem has taken over the function of photosynthesis from the leaves, either owing to the reduction of the latter to mere scales or to their complete disappearance. Where *chlorenchyma*, i.e., chloroplast-containing tissue, is present, whether of the palisade form or not, there will usually be some water storage tissue closely adjacent to it in the cortex.

The cells of such a tissue are large and thin-walled, and they not infrequently contain mucilage. An it is nearly always in xerophytes that chlorenchyma occurs in the cortex the advantage of a water reservoir closc at hand is obvious. Indeed, in the Cacti, where the stele is often relatively slender, the greater part of the stem tissue is water-storing cortex, only a narrow outer zone being chlorenchymatous. Internal glands and sclereid cells are frequently present in the cortex, the latter being sometimes complexly branched and forming an important part of the skeletal system of the plants (e.g., magnoliaceae) in which they occur. More often, however, they are isolated and scattered and not of significant mechanical value. After elongation of the young stem is over, a ring of sclerenchymatous fibres may form in the cortex, as for example, in Oak, Birch or Ash, supplementing the collenchyma as a strengthening tissue. Growth in girth soon and repeatedly ruptures this ring, but the gaps are made good by the ingrowth of parenchyma cells into the radial spaces due to the ruptures, and their transformation into sclereids. Thus a composite mechanical ring is formed of arcs of fibre cells separated by arcs of sclereids.

One of the most striking departures from normal cortex formation is to be seen in aquatics, where the cortex, or in the case of Monocotyledons the whole ground tissue, is divided into large air spaces, separated by membranes only one the cell thick. These lacunae are limited in length to one internode in Dicotyledons stems, and even in Monocotyledons they are not of great vertical length, but the horizontal membranes which separate them are perforated by minute intercellular passages, which form at all the angles of the cells. These provide for the movement of air from one lacuna to another, but they are too small to allow water to penetrate, if it should chance to get into one lacuna through an injury to the epidermis. They are in effect waterproof bulkheads and prevent the accidental flooding of the whole internal air system.

PERIDERM

The epidermis is usually only retained for a short time while the shoot is young, and it is later replaced by a secondary covering. This is necessary because the thick walls and the cuticular layer of the epidermis inhibit the division of its cells. The epidermis can therefore only accommodate itself to the increasing girth of growing stems by the tangential stretching of its cells, and that is limited in extent. A few cases exist of woody plants which retain their epidermis for a number of years, e.g., *Laurus*, *Aucuba*, *Rosa*, *Acer*. The cuticle is soon ruptured by the growth of the stem, but the cutinized cell wall increases enormously in thickness and is progressively regenerated from within, while the outer layers crack and crumble away. Most smooth, stemmed shribs do, however, eventually form a bark.

One exception is the Mistletoe, in which the primary epidermis is supplemented by further cutinized layers of cells formed successively inwards, so that a multiple epidermis of great thickness is build up. Only a growing tissue can retain permanently its equilibrium with other growing tissues, and thus the secondary covering of the stem is initiated by the formation of a cambial layer known as the *cork cambiun* or *phellogen* the chief product of which is the *phellem* or *cork*. The

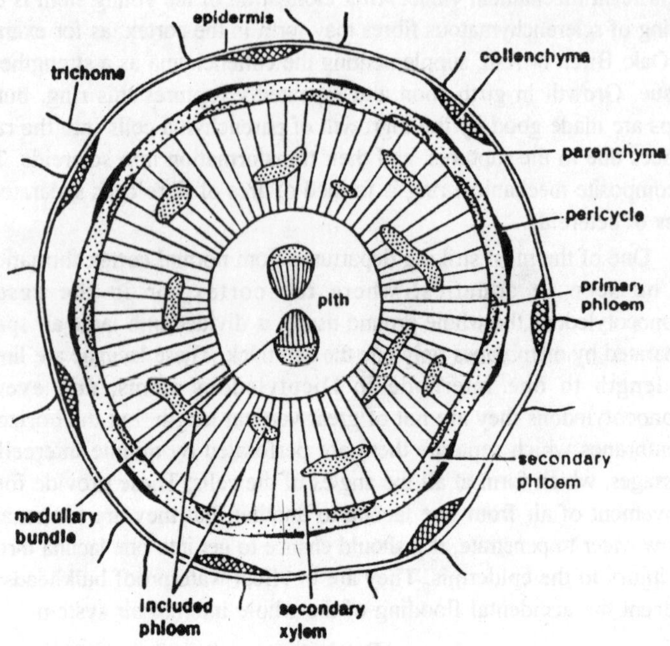

Fig. 1.10. T.S. of stem of Achyranthus.

phellogen is a true secondary cambium formed by the rejuvenation of tissues already fully differentiated, which may be situated anywhere between the secondary phloem and the epidermis itself, although in most cases the first phellogen lies in the cortex near to the outer surface. It is a single-layered meristem, and by tangential divisions it cuts off a succession of cells outwards, which are arranged in radial rows, corresponding to the cells of the phellogen from which they have arisen. There are a few examples of intraxylary phellogens, e.g., *Ariemisia*, where cork formation begins in the wood of the current year, forming a sleeve over the wood and ending upwards in a dome in the pith near the stem apex, thus isolating all new growth from the old. This may be related to the desert conditions under which the plants concerned chiefly live.

Priestley has shown an association betwen phellogen formation and the existence of some anatomical barrier to outward sap diffusion. When a functional endodermis is present this acts as such a barrier and the phellogen commences in the pericycle. In the absence of such an endodermis there may be no corresponding barrier until the hypodermis or the epidermis is reached, and it there that phellogen appears. Intermediate cases may be accounted for by the presence of an internal cuticle at some level in the cortex or by the existence of an impermeable sclerotic layer. According to Priestley's view the accumulation of sap on the inner side of a diffusion barrier provides the physiological stimulus to increased cell division. As the cell sap always contains diffusible fatty substances the materials for the formation of suberin in the cork layers are also accumulated at the level of the phellogen.

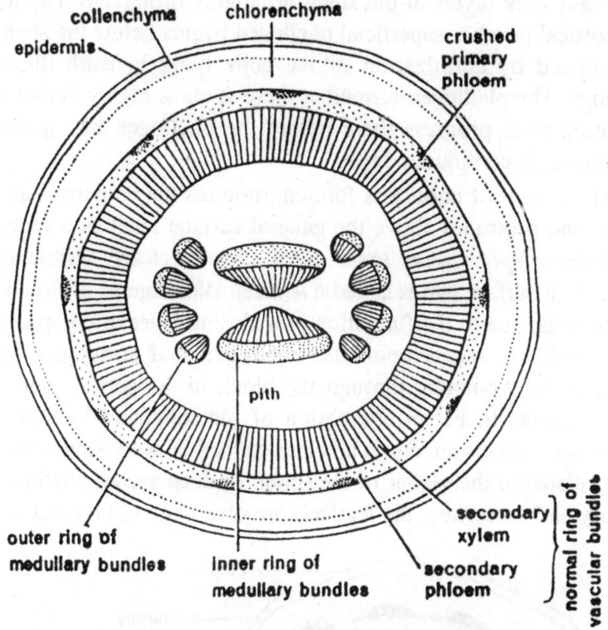

Fig. 1.11. T.S. of stem of Boerhaavia.

The suberization of the walls of phellem cells consists in the deposit of a layer of suberin, unmixed with cellulose, between the middle lamella and the inner wall of pure cellulose. This layer varies greatly in thickness in different cases, and in the massive "cork wings" on the branches of species of some trees, such as Ulmus and *Acer*, it may be practically absent, which has led to the application of the term *phelloid* to such unsuberized cork tissues. The suberin itself is apparently a mixture of the anhydrides of fatty acids, the chief of which is *pehllonic acid*, $C_{22}H_{43}O_{30}$ and it is thus closely similar to cutin. Both substances seen to be dependent on contact with air for their development.

There are no pits through the suberin layer, and as it is highly impermeable the suberized cell soon dies. Tannin and related substances often accumulate in the cells before suberization, and these give the cork its dark colour and its value for tanning leather. Haberlandt showed that the protective value of two-year-old cork layers against evaporation is about the same as that of the primary epidermis. That it is not greater is probably due to the cracking, caused by growth pressure, which fissures the outer layers of the cork, and would destroy its protective value if fresh layers were not constantly added from within by the phellogen. The heavy cork layers of old stems are highly protective. The formation of a cortical or other superficial phellogen begins below the stomata, and is preceded by cutinization of the cells lying beneath the stomatal openings. The phellogen formed at these spots is highly active and cork formation is in progress there before the phellogen has spread round the stem as a continuous layer.

The excess of cork cells formed, ruptures the epidermis around the stoma, and protrudes above the general surface level as a spongy mass called the *complementary tissue*. Such a spot is clearly visible to the eye on the stem surface and is called a *lenticel*. Blockage of diffusion through the lenticels, due to the formation of a closing layer of compact phellem in the autumn, reacts upon the phellogen and stimulates increased divisions, thus bursting through the block in the spring. The life of a lenticel seems to be an alternation of blocking and bursting phases. Permanent lenticels are not in all cases formed at every stoma, the number being related to the vigour of the shoot. In deep-seated phellogens, both primary and secondary, the lenticels are obviously not related to stomata

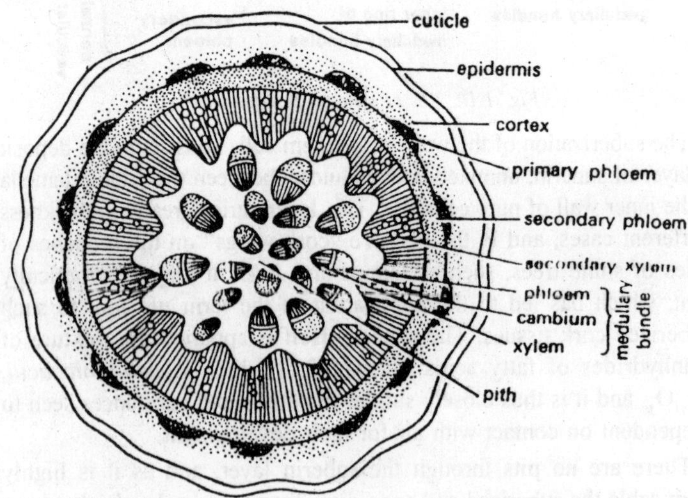

Fig. 1.12. T.S. of stem of Bougainvillea.

and their distribution appears to be related rather to the broad medullary rays which are associated with the leaf traces. Lenticels have been interpreted as ventilating apertures, but their causation has nothing to do with the assumed necessity for this function.

Suberin is, however, exceedingly impermeable to air, and the rounded and irregularly arranged cells at the lenticel undoubtedly do allow access to the underlying tissues for air, which cannot penetrate at any other point. The physiological value of this is reflected in the bad effects on the health of the plant which result if the lenticels become covered with external growths of moss or lichens. While the chief activity of the phellogen is in forming phellem outwards, there is usually also a slow production of cells inwardly, which are added to the cortex and are known as *phelloderm* or secondary cortex. These cells are not suberized and are only distinguishable from the cortical parenchyma by their arrangement in radial files and sometimes by the presence of starch grains. Their walls are often thicker than those of the cells of the primary cortex and may even be lignified, as in *Canelia*. We thus see the elaboration of a secondary protective system consisting of phellem, phellogen and phelloderm which goes by the collective name of *periderm*. This slowly develops into the *bark* of old stems, but in this development other tissues become involved. In smooth-barked trees the first phellogen remains active for some years and is only slowly replaced, if at all.

The bark of such trees therefore consists of a continuous cork film. In rough-barked trees, however, the first phellogen is soon replaced by a series of others, lying successively deeper in the tissues, and all later phellogens, after the first few, are formed in the secondary phloem, much of which is destroyed in this way and its remains incorporated in the bark, along with dead cortex, collenchyma, sclerenchyma, etc. Bark is thus a composite structure. Where the later phellogens form continuous zones the bark may show successive annual rings, like the annual rings of the wood, each marking the formation of a new phellogen, but in many other cases the later phellogens form a series of discontinuous arcs and the bark is thus cut up into distinct areas, between which deep cracks appear.

The various bark patterns on old tree trunks which are characteristic of different species, are thus traceable to variation in the arrangement of the phellogen layers. An abnormal formation of cork is commonly associated with injuries, when it is known as *wound cork*. Wounding or almost any sort of lesion of the integrity of the plant body may call it forth, and any living tissue may produce it. The first reaction to the wound is the suberization of the outer surfaces of exposed but intact cells. Suberization then spreads inwards for several cell layers. Below this barrier cell division starts and a cambium is organized which produces a few layers of regular cork cells, forming and impermeable barrier against desiccation, parasites and other inimical influences.

ENDODERMIS

The innermost layer of the cortex, surrounding the stele, is called, as in roots, the *endodermis*. A typical endodermis, with Casparian bands, recognizably the same as that in roots, is not common in stems. The chief exceptions are among water plants, such as *potamogeton* and *Hippuris*, in which the vascular tissues are concentrated into a relatively slender axial strand, as is normal in roots. Many other exceptions occur which are not, however, related to special environmental features, e.g., in some Compositae, species of *Primula*, and in the rhizomes of some Monocotyledons. The endodermis, in most cases where it occurs in the stem, is a common investment round the stele as a whole, but there are a number of instances among Angiosperms where each vascular bundle has an individual endodermis, which is not infrequently continued into the leaves. Species even within the same genus may differ in this character as, for example, in *Ranunculus*, where *R. lingua* and *R. flammula* are peculiar in having individualized bundle sheaths.

The significance of the endodermis and of its variations cannot as yet be fully interpreted on physiological grounds, but some interesting points have been elucidated. It first appears outside the procambial zone at the stem apex, simultaneously with the first appearance of xylem elements. Materials from the breakdown of the xylem protoplasm probably diffuse out radially, and may supply the fatty acids which condense to form the Casparian Band. It has been suggested that the condensation of fatty acids at this level is due to the penetration of air to this zone through the intercellular spaces of the cortex, but while we may have here the foundation for an explanation it is plainly not the whole story. The stele is not the only organ of the stem to possess a sheath, since many glandular passages are similarly protected, nor does the stelar sheath always have a true endodermal character.

In a great many stems the layer surrounding the stele is marked by the presence of large starch grains, although the cells may not be otherwise different from those of the rest of the cortex, which may contain plastids but no large starch grains. This "starch sheath" corresponds morphologically to an endodermis, though it will avoid confusion if we do not apply this term to it, since it lacks the Casparian thickening and is physiologically different. Its morphological identity with an endodermis is shown by the fact that a starch sheath in a young stem may sometimes lose its starch and become thickened as an endodermis as a later stage.

Cases also occur where the starch sheath is dissected into separate arcs or strands, divided by ordinary parenchyma, as in the Nettle and at the nodes of Grasses. These large starch grains are movable, sinking always to the lowest side of the cell vacuole, and they have been called *statoliths*, and assigned a function in connection with geotropism. In underground stems of Monocoty-ledons, such as *Convallaria*, there is a pronounced endodermis, which may be double or even triple. In the aerial stems, however, the boundary of the stele is usually a sclerotic or collenchymatous zone which grades insensibly into the cortex. It is evident that the stelar sheath in the steam is by no means uniform, and Strasburger has coined the useful name *phloeoterma* to apply to the innermost layer of the cortex, which must be recognized to be an important anatomical boundary, whether or not it has the character of a true endodermis.

Primary Vascular Tissues

Everything within the endodermis is denominated the *stele.* The question of the applicability of this concept, derived from the study of the stem in the Vascular Cryptogams, to the stems of the Spermatophyta, we have discussed previously. It is at any rate certain that in the absence of an endodermis, and particularly in some Monocotyledons, it is impossible to give it the precise connotation which it has, for example, in the *Ferns*; but it is a handy descriptive term and we shall continue to apply it, without prejudice, to the complex of tissues lying within the cortex and of which the vascular tissues form so important a part. Its outer boundary is usually called the *pericycle.*

The pericycle in the stem is very different from the meristematic zone which goes by the same name in the root. The idea of a pericycle is that of a cell layer formed from the outer zone of the procambium, within the endodermis and immediately surrounding the phloem. This position in the stems of Dicotyledons is often occupied by a zone of sclerotic fibres or of sclereids, sometimes continuous, sometimes broken up into groups which lie outside the vascular bundles. At the present day it is doubtful whether this fibrous zone is really distinct in all cases from the phloem, for it has been shown that it may arise from protophloem in which the sieve tubes and other soft elements have been obliterated and only phloem fibres remain. Whether this is always so cannot yet be stated. The vascular bundles of Monocotyledons, whether they have an

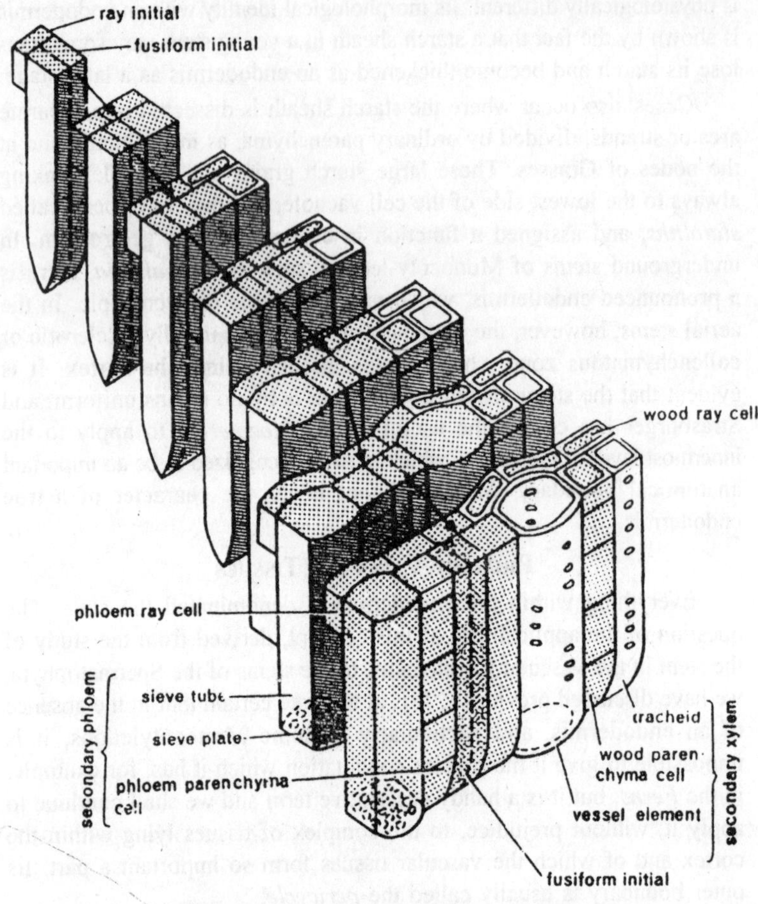

Fig. 1.13. Figure showing vascular elements.

individual endodermis or not, are frequently surrounded by sclerotic sheaths, complete or partial, which are quite independent of the phloem but are otherwise analogous to the general pericyclic fibre-sheath in Dicotyledons. It is difficult to see why both types of sheath should not be classed under the same name, as the bundle sheaths in Monocotyledons are particularly characteristic of those stems in which no general pericycle exists.

Each *vascular bundle* is a complex of tissues, partly conducting elements, partly storage elements and partly strengthening elements. Haberlandt gives the following analysis of the components: While *xylem* and *phloem* denote the whole of the two main tissue systems in the bundle, Haberlandt applies the terms *hadrome* and *leptome* to the two sets of conducting elements respectively, to distinguish them from the fibres present. The most general arrangement of tissues in the bundle is the *collateral*, in which both tissues lie on the same radius, the phloem constituting approximately the outer half of each bundle and the xylem the inner half. We have described above the first appearance of protoxylem at the inner side of the procambial strand and of protophloem at the outer side.

The further differentiation of each tissue from procambial cells is centrifugal in the case of the xylem and centripetal in the phloem. In Monocotyledons generally, and in some reduced herbaceous types among the Dicotyledons, the whole of the procambium is thus differentiated and no further addition of vascular tissue is possible. Such bundles are called "closed." Among Dicotyledons, however, it is usual for a zone of cells about half-way across the procambium to become *cambium*, which remains meristematic and thereafter adds cells continuously to both xylem and phloem in radial files. This change marks the beginning of what is called *secondary thickening*. As we have previously pointed out, the organization of cambium may begin so early that every part of the vascular tissue is formed from it and the radial arrangement rules from the beginning. Such cases show that a clear distinction of primary, or procambial growth, from secondary, or cambial growth, is not always possible.

Even in plants with a meristematic ring which gives rise to a woody stem, differentiation normally begins at a number of separate points on the ring, so that separate primary bundles are for a short time distinguishable. The few exceptions, such as *Vinca*, in which a continuous ring of xylem is present from the beginning, may really be analyzed into a ring consisting of a very large number of uniseriate bundles, placed very close together and showing a radial arrangement of the elements from the start. The residual meristem between the original bundles may organize directly into *interfascicular cambium*, continuous with that in the bundles, or it may differentiate into parenchyma.

In the first case secondary xylem and phloem are differentiated in the spaces between the original bundles and secondary growth goes forward both in the bundles and between them, forming continuous zones of xylem and of phloem, as in timber trees. In the second case a layer of the interfascicular parenchyma may later differentiate into cambium, but such secondary cambium forms only or mainly parenchyma, and the original bundles remain permanently distinct, i.e., secondary vascular

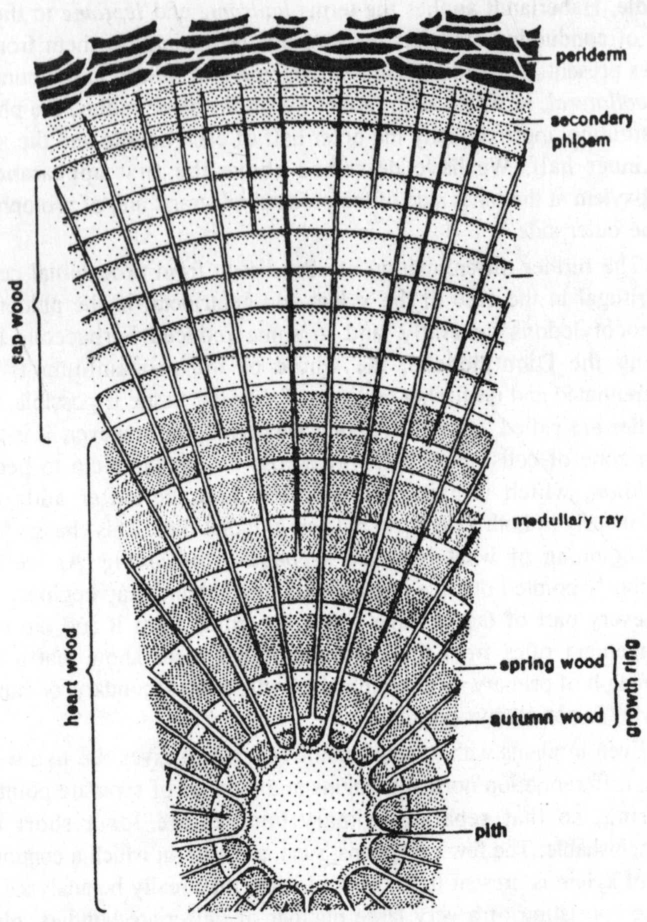

Fig. 1.14. Section showing heart wood and sap wood.

tissue is confined to the bundles. This is not uncommon in woody climbers like the Vine. On the other hand, the spaces between the bundles may remain permanently parenchymatous, with no interfascicular cambium at all, and secondary growth of the bundles themselves is very limited. This is the extreme herbaceous type of stem. The parenchymatous intervals between the original bundles are the *primary medullary rays*. They are usually broad at first, and in stems of the Vine type they grow radially by additions from the interfascicular cambiums, so that they retain their initial breadth permanently, but in timber trees the original broad rays are soon bridged and closed by the interfascicular xylem.

Certain interfascicular cambial cells, however, produce only parencyma cells, both inwards and outwards, so that at these places narrow rays are built up, which may be uniseriate, or at most two to three cells broad. These narrow rays continue outwards through the whole zone of vascular tissue, no matter how broad it may become, from the pith to the pericycle, interrupted only by the cambium itself. Medullary rays which originate after secondary thickening has begun are called *secondary rays*. Their point of origin is an initial parenchymatous cell, cut off, usually terminally, from one of the fusiform cambial cells. Similar narrow rays are formed in the original bundles as soon as radial growth from a cambium is established, and they may be similarly permanent. The details of secondary growth we will leave until later, while we consider some characters of the individual bundles.

Two types besides the simple collateral bundle deserve mention. The first is the *bicollateral*, in which there is a second phloem region, usually smaller than the first, inside the xylem zone. This internal or *medullary phloem* may be confined to the primary bundles, or it may develop in the interfascicular region as well, so that a more or less continuous zone of internal phloem results. It is a character found only in certain families (e.g., Cucurbitaceae and Solanaceae), sometimes in members with all types of growth, sometimes only in climbing species of the family. The second type is the *centric bundle*. This is usually, though not invariably, a closed bundle and is commoner in Monocotyledons, especially in the underground stems, than in Dicotyledons. It may have the xylem in the centre and the phloem outside in which case it is called *periphloic* or *amphipholic*, or else the reverse when it becomes perixylic or amphixylic.

The closed bundle which is characteristic of the aerial stems in Monocotyledons is small and has relatively few elements. It is sometimes described as the Y-type from the characteristic grouping of the xylem elements. The protoxylem, which is often crushed and obsolete at maturity, occupies the tail of the Y, and the arms are formed by, or end in, two very large vessels, between which lies a group of small lignified tracheids. Above this is the compact phloem, consisting of regularly arranged sieve-tubes and companion cells with a few parenchyma cells at the sides.

Another monocotyledonous type has a still simpler xylem, formed of one or two narrow protoxylem elements and one extremely large vessel in the centre of the bundle. This type of bundle is formed mostly in monaxial Monocotyledons with very large, rapidly growing leaves, e.g., Banana, and the extra large vessels are probably correlated with the need for quick development of the water-conducting capacity of the stem. The concentric type of bundle is considered to the most advanced type in Monocotyledons and has probably been derived from one of the collateral types. As is well known, the arrangement of the vascular bundles in Dicotyledons is typically in a single, wide ring, while in Monocotyledons they are typically dispersed, seemingly at random, across the transverse section of the stem.

In the Dicotyledon the orientation of the vascular tissues is constantly with the protoxylem innermost but in Monocotyledons, although the general tendency is the same, there is less uniformity, especially in the bundles near the centre. The parenchymatous tissue enclosed by the ring of bundles in Dicotyledons is the *medulla* or *pith*, but in monocotyledonous stems there is no corresponding region, and the parenchyma among the bundles is generally known as the *ground tissue* or *conjunctive tissue*. Departures from these typical arrangements will be dealt with later. We must now consider the bundle systems in three dimensions in order to get some idea of vascular architecture.

In Dicotyledons the leaves have usually a narrow base of insertion on the stem. The number of *leaf traces* is therefore nearly always small and may be only one. These traces differentiate downwards from the leaf base and intercalate themselves between the traces of lower leaves. In this way the trace of a young leaf may reach downwards through several internodes. Eventually, however, at a node, it either becomes united laterally to the adjacent bundle in the ring, or it forks and unites to the bundles on each side of it, producing "synthetic" bundles. The distance it reaches downwards before this happens depends on the leaf arrangement on the stem, since it appears that the traces to which a given leaf unites itself are those of the leaf vertically below it is in the leaf spiral.

At every node some anastomosis of traces takes place, so that there is at these levels a considerable amount of vascular linkage, while in the internodes the bundles remain separate. The lateral traces of the leaf, where such exist, are close to the median trace at the leaf base, but as they go downwards they are separated by the intercalation of traces from higher leaves, until they may lie far apart. Where there are several traces to each leaf they will thus come to be spaced out where they enter the ring, and in types with many traces, such as *Liriodendron*, they may occupy points all round the ring before they finally join the ring bundles. The lateral traces join the ring at a higher level than the median trace, which means, in terms of apical development, that they do not differentiate until several plastochrons later than the median trace, which is always the strongest and longest and goes farthest down the stem. While the trace differentiates downwards, the period of growth available for differentiation gets shorter in each successive internode and the trace will therefore contain less and less primary xylem and more and more secondary xylem, until it is finally merged in the general mass of secondary xylem.

The phloem generally accompanies the xylem, though peripheral portions of phloem may branch off independently and join up with others far above the level at which as a whole finally joins a synthetic bundle. Comparative studies give good grounds for the opinion that the primitive leaf trace system in Dicotyledons is one with three trace bundles leaving three separate gaps in the vascular ring of the stem, the so-called *trilacunar* type. Reduction to one is due to some cases to abortion of the lateral traces, in others to fusion into one, which, however, may be three-lobed at the base. The many-bundled type of trace is also derivative from the three-trace type, which is frequently found in the seedlings and youngest leaves of plants with complex traces.

Even in Monocotyledons the three-trace type is found in many small and slender species and in the young leaves of others, and it appears to be primitive in this group also. In Monocotyledons the leaf base is broad and surrounds the stem, even in the embryonic state. It may contain thirty to forty trace bundles, which enter the stem simultaneously all round the periphery. Each leaf primordium is quickly pushed outwards away from the apex by the new encircling primordia which follow it, with the consequence that its traces become acutely fixed at the node, bending outwards from a central position into the leaf base.

In Monocotyledons with extended internodes the bundles follow a course downwards which is similar to that in Dicotyledons, tending gradually outwards towards the point of anastomosis with older bundles, which appear to be usually those of the next leaf vertically below the leaf considered, i.e., on the same orthostichy. At the nodes the number of anastomosing bundles is so great that it creates a nodal plexus of intermingled bundles which may extend right across the stem, and is increased by the insertion of numerous traces from the axillary buds. It must be admitted that the course of the bundles in even a simple Monocotyledons is not yet known with sufficient accuracy to warrant positive statements, while in those species with contracted stems the complexity and irregularily defines analysis. The whole question of vascular development is indeed in a somewhat hazy state. For example, although basipetal differentiation of vascular elements within the procambium seems to be the rule, it is by no means certain that the differentiation of the procambium itself, from the primary meristem, is not acropetal or continuous.

The above account must therefore be regarded as provisional. The development of the meristematic cells in a young leaf axil into an axillary bud complicates the dicotyledonous nodal structure further by its special system of traces. The axillary bud has an apex which is a miniature of the main apex and has the same procambial arrangement, giving rise to a duplicate of the stem stele. As this differentiates downwards into the stem, the ring or group of traces opens out fanwise, forming two bundles which pass into the main ring between the bundles which flank the median leaf gap and join them above the level at which the median trace of the subtending leaf itself joins an older trace in the ring. Exogenous adventitious buds such as those on the corms of *Cyclamen* are apparently connected secondarily to the stele of the parent stem. Procambial strands come from the bud and differentiate inwards through the cortex towards the stele, and the formation of vascular tissue in this procambium is likewise downwards from the bud.

Cambium

It has been previously pointed out that the organization of cambium from procambium marks the beginning of radial growth, commonly called secondary growth, which is capable of indefinite expansion outwards and will, if continued, build up a massive woody stem. In many herbaceous stems, although secondary growth begins as in woody plants, it does not continue beyond one year. The structure of such herbaceous stems is therefore similar to that in the one-year old stems of woody plants. The *Cambial initials* are distinguished in the procambium by their repeated tangential, longitudinal divisions, giving rise to a zone of cells which appear narrow in transverse section. This occurs in the part of the stem apex where active growth is still going on, and the cambial initials undergo considerable elongation and become radially compressed by the expanding tissues of the pith and cortex.

In spite of these changes the cambial cell remains a true meristematic cell, and its repeated tangential divisions cut off new xylem cells centripetally and new phloem cells centrifugally. It is difficult to distinguish these new tissue elements in their early stages of development from the true cambial cells, and it is not possible to say in every case whether the fully developed cambium consists of only one layer of cells or of more. The position at which the cambium forms in the procambium

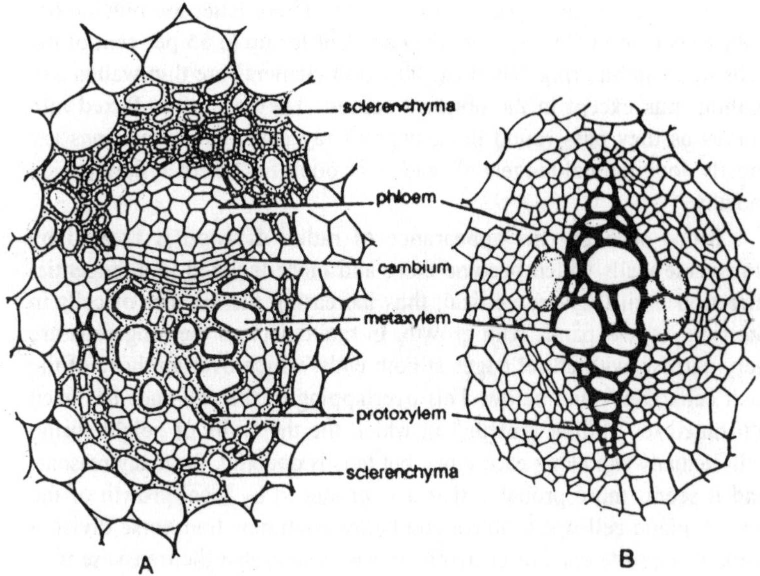

Fig. 1.15. Vascular bundles. A—endarch; B—exarch.

may be physically determined by the gradient between acid sap from the differentiating wood and alkaline sap from the phloem. This theory, which is due to Priestley, maintains that the cambial layer lies at the pH on the gradient which corresponds to the mean isoelectric point of the cell proteins and that the protoplasm of the cambial cells remains unvacuolated because at the isoelectric point colloids have their minimum affinity for water and their maximum density.

The same theory extends to the formation of interfascicular cambium from fully differentiated parenchyma cells and perhaps also to the appearance of the phellogen. Both these charges involve the alteration of cells with large vacuoles and little protoplasm into meristematic cells which have no vacuoles and abundant protoplasm. The isoelectric point of the cell proteins is the condition which would seem most favourable to such a change. The cambium does in fact lie between acid xylem (pH 3.4 to 5.0) and alkaline phloem (pH 7.8), while the phellogen lies between the strongly acid cork (pH 3.0) and the less acid cortex (pH 5.5 to 6.5). Protoplasm at its isoelectric point will lose water to vacuolating cells and the condensed non-vacuolate condition produced is that most favourable to protein synthesis. When cambium is viewed in tangential longitudinal section, two types of cell are seen, the long, pointed fusiform cells which give rise to the vascular elements, and the small, isodiametric cells in groups, which produce the medullary rays. These latter are much more numerous than in Conifers and may account for up to 55 per cent of the cells in a cambial ring. The long, fusiform elements are thin walled and without pits except in the dormant season. They are often flexed into curves as they pass round the groups of ray cells. Their divisions are mostly longitudinal-tangential, and it is doubtful whether they divide radially.

More probably the appearance of radial division is created by transverse walls which become more and more inclined in a tangential direction until they are vertical, thus increasing the number of cells in the ring, as it expands with growth. In radial section the long cells are very narrow, with chisel edges at both ends, which overlap those of the next cells above and below. This overlapping of the cell ends has been attributed to "sliding growth," in which the the walls of neighbouring cells actually slip along each other, but thus is doubtful, for many reasons, and it seems more probable that it is produced by local growth of the young, platic cell-walls, above and below each new transverse division wall, at opposite ends respectively, in such a way that the transvese wall is tipped into the vertical without any mutual slipping of the cells. The cambial ray cell, on the contrary, divide in all three planes. Bailey has shown that in Dicotyledons there are three main types of cambium, corresponding to three types of wood structure.

1. Initial cells averaging 890 microns in length, producing long, narrow vessel segments with scalariform pitting of the end walls.
2. Initial cells averaging 410 microns, producing short, broad vessel segments with cross walls which are simply pitted or else absent.
3. Initial cells averaging 250 microns, producing vessels similar to type 2, but the cambial cells are arranged in regular horizontal rows (Etagen cambium) and give rise to a corresponding "storied" structure in the wood.

These three types are regarded as marking advancement in specialization of the vascular structure. The cambium in the resting state, in winter, is sharply marked off from the vascular tissues, with fully formed vessels and sieve tubes lying in immediate contact with the cambial cells from which they have been formed.

During the summer activity, however, the cambial zone seems much broader and less clearly delimite, because tangential divisons follow one another so rapidly that the true cambial zone is flanked on both sides by a layer of still undifferentiated cells, and the transition to the mature vascular elements is thus spread over a broader zone of development. It has been shown that the cambial growth in spring begins in the swelling buds and spreads downwards from them. It begins simultaneously all over the tree, but the rate of spread is much slower in the older branches and its stops sooner in them than in the young shoots. There is good evidence that the stimulus to cambial growth is the diffusion of hormones from the leaf initials in the bud. Auxin-α and heterauxin can both start strong cambial growth when they are supplied to a shoot, and auxin has been found in naturally active cambium and is known to be produced by leaf initials.This direct relation of leaf development to cambial activity is probably the reason for the proportionality between the extent of leaf surface and the amount of vascular tissue formed in the stem, which observation has revealed.

The hormone explanation may hold good not only in growth of the normal cambium but in the differentiation of new vascular tissue from old permanent tissues, which occurs in regeneration growth after wounding. Such new developments show a definite directional guidance from point to point, often around obstacles, which strongly suggests the action of a diffusing hormone. This is also probably true of new, adventitious, vascular differentiation, as in the case of *Cyclamen.* The cambium, like other meristematic tissues, does not itself undergo any marked change throughout its existence. The only visible indication of age is a gradual increase of the average length of the cambial cells and of the tracheids formed from them, an increase which may continue for a number of years from the beginning of cambial growth.

Eventually, however, a steady state is reached and no further change in length occurs. In the most pronouncedly herbaceous plants there may be no interfascicular cambium, and the separate bundles themselves have little or no cambial growth, while in the most extreme herbaceous types there is not even a continuous ring of procambium. Indeed the annual, monocapric habit of growth, limited to a single growing season, is probably brought about by the absence of early disappearance of cambium and the consequent lack of the power of extended growth. Monocotyledons are generally supposed to have no cambium in their bundles, but the metaxylem elements are often in radial rows, which is characteristic of cambial growth, and they probably resemble the annual Dicotyledons in having cambia that are very short lived.

The specialized cambia of some Monocotyledons which undergo secondary thickening are dealt with. In connection with the cambium we may briefly mention the abnormal tissue formation known as *callus*, which is associated with wounds and similar injures. The name is applied to irregular tumour-like outgrowths, due to a resumption of meristematic activity called out by the injury. Any living tissue may contribute to this development, but where the cambium is involved, its contribution is much the greatest. Thus a callus ring arises on the cut ends of branches or old roots, which is almost entirely due to growth from the exposed end of the cambium and which spreads until the wounded surface is completely covered. This new tissue is almost entirely parenchymatous, and there is no regular differentiation of tissues, as in normal cambium growth, At the most a few irregular scattered tracheids or sclereids may be formed, embedded in the mass, while the outermost cells may become corky.

Secondary Vascular Tissue—Phloem

The *metaphloem*, including the secondary phloem, if any, consists of sieve tubes, their companion cells and phloem, parenchyma. The companion cells, originally sister cells of the sieve tube cells, usually divide transversely more than once, so that they are individually much shorter than the sieve tube segments. The phloem parenchyma cells are variously shaped, and may be so long and narrow that they scarcely justify their name. Mingled with there elements in many plants are phloem fibres, either scattered or, as in *Tilia*, in tangential bands alternating with the conducting tissues. These fibres closely resemble those which form the fibres of the so-called pericycle, external to the phloem, and are often classed with them as *libriform fibres*, from the old name "liber" (on "bast") applied to the inner bark, in which the phloem was distinguished as "soft bast" and the fibres ad "hard bast."

Fibres of the secondary phloem, being produced from the cambium, show the same radial arrangement as other phloem cells, but the pericycle fibres, coming direct from procambial cells, do not. Libriform fibres are narrow and very highly thickened, and their pits are very small, so that the pit opening shows little or no border. The pit openings are slit-like and the openings on the inside and the outside of the wall are usually at right angles to each other, so that the pits appear like minute crosses on the cell wall. They are longer than the xylem fibres, but flexible and tough rather than brittle. Jute, and several other important fibres consist of this type of cell.

Secondary Vascular Tissue—Xylem

The term *metaxylem* includes all the secondary xylem. The cell structrue of secondary xylem is very variable, and each type of wood has its recognizable characteristics, by which it may be identified. The usual constituents are vessels or tracheae, tracheids, xylem parenchyma and fibres of various types. In true vessels the successive elements open

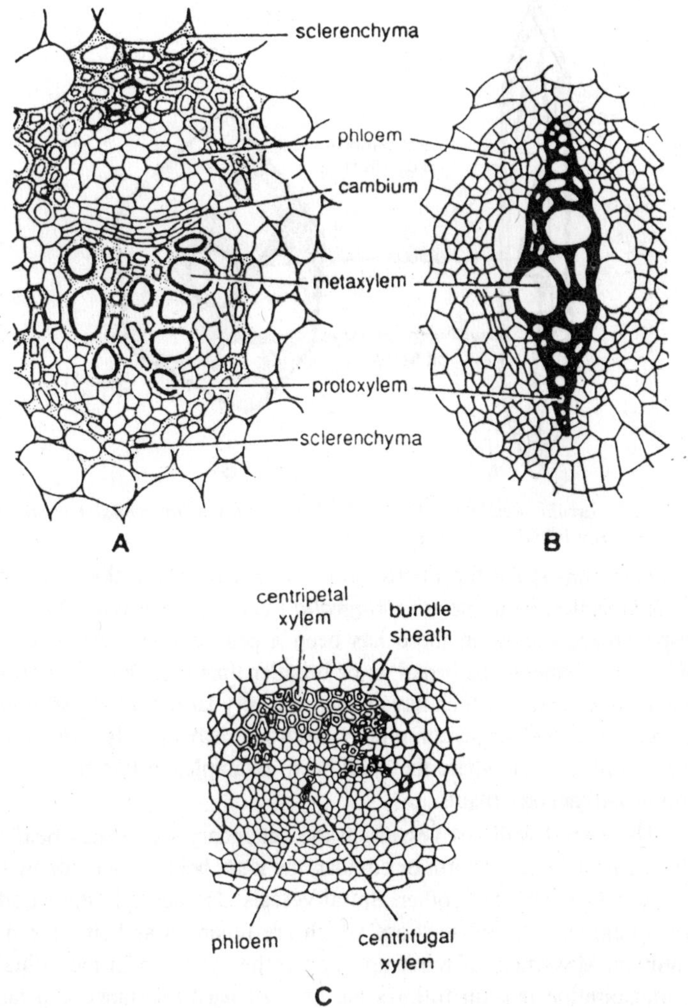

Fig. 1.16. Position of protoxylem in vascular bundles. A—endarch; B—exarch; C—mesarch.

into one another in longitudinal succession by membraneless portals so that the lumen forms a continuous passage. This involves the alteration or disappearance of the transvese walls which separate the young elements from each other. Comparative studies have shown that in the wood of certain families which are predominantly arboreal in habit the vessel elements are narrow and long and have steeply oblique end walls

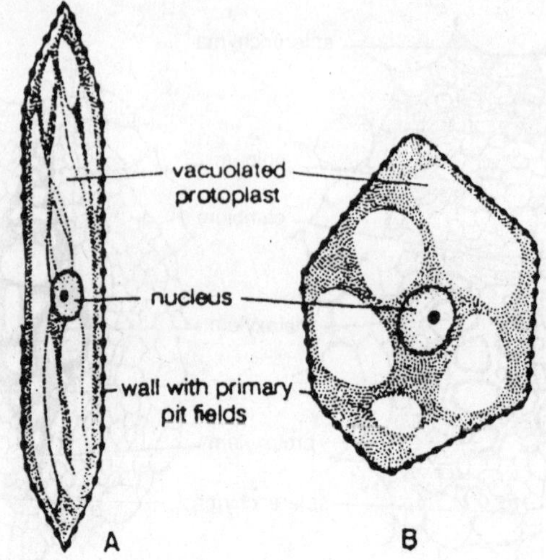

Fig. 1.17. Vascular cambium: A—two fusiform initials in tangential section; B—a ray initial in tangential section.

with numerous scalariform perforations. In this they resemble the vessels of the Gnetales. Example are : *Magrolia Betula, Salix Myrica.* From this unspecialized condition there has been a progression towards vessels with short elements, as broad as or broader than their length, and with transverse end walls which have either one very large central perforation, or are, in the final stage, completely absorbed, leaving only a ring on the vessel wall to mark where they had been. Examples of this are common among herbaceous plants.

The lateral walls of vessels in the secondary wood may bear pits which are either scalariform or circular, the latter being either side by side or alternating with each other. Spiral vessels also occur in the wood of many families, but not in those which are regarded as being the most primitive. Movement of water upwards in the xylem under the influence of transpiration tension follows the lines of least resistance and large, fully open tubes must obviously afford an easier passage than narrow elements with many cross walls, even if the latter are perforated. Climbing plants which have very long, narrow stems are those in which the resistance to hydraulic flow is greatest, and they are notable for the large average diameter of their vessels, Poiscuille's Law of fluid flow expresses the velocity thus:

$$V = \frac{\pi p r^4}{8 \ell \eta}$$

where, V = velocity of flow

p = pressure of difference.

r = radius of tube.

l = length of tube

η = coefficient of fluid viscotiy.

The velocity is directly proportional to the fourth power of the radius and inversely proportional to the length of the path of flow. Velocity is thus such more sensitive to differences of radius than of length, and the resistance offered by a long stem can be readily compensated by the increased width of the vessels. To double the radius of a vessel will increase the flow capacity by sixteen times, but to halves the length of the path will only increase the flow twice.

Owing to the irregularites of their structure the vessels may have conducting capacity which is no more than 30 to 40 per cent of the theoretical efficiency but their comparative capacity is nevertheless governed by the hydraulic laws. The flow of water upwards proceeds through very numerous fine channels, branching and rejoining which are formed of single vessels or or small groups of vessels islanded among tracts of non-conducting parenchyma and fibre cells. Among herbaceous plants the leaf trace bundles serve for conduction only to the leaves to which they are connected, and there is some evidence that this physiological segregation may persist downwards in the synthetic bundles of the lower stem. Whether this applies to the massive secondary wood in trees is doubtful. There is no positive evidence of the anatomical segregation in the mass of wood of distinct conducting tracts associated with individual branches.

The selective flow through limited channels, which may be observed when a single leaf or branch draws a coloured solution through the main stem, is probably conditioned by the downward transmission of the water tension, which is the motive force of the flow, through the physically most direct path available. Vessels anastomose freely both tangentially and radically, so that the water currents, even if confined to the vessels, may shift their paths readily in either direction, even from one annual ring to another. The force of transpiration is not, however, sufficient to draw water across the medullary rays, in the absence of a vascular connection. In most woody plants the vessels only remain functional for, at most, a few years after their formation, and frequently only for one year. The outer most annual rings of wood are often lighter in colour and softer in texture than the older wood. They form the *alburnum* or sap wood, while the older rings compose the *duramen* or heart wood.

Only the outermost portions of the sap wood may be active in conducting water and the heart wood takes no part whatever in conduction.

Water storage in the duramen is confined to the living cells of the medullary rays and xylem parenchyma, but as the conducting elements pass out of action they are often filled with resinous or gummy substances and tannins, and the walls may be impregnated with colouring matters, an example of which is Haematoxylin, which oxidizes to the well-known dye, Logwood. Minerals substances, such as Calcium carbonate in *Fagus* and *Ulmus*, Silica in *Tectona grandis* (Teak) or Calcium oxalate in Chloroxylon (satinwood) may also impregnate the heart wood, adding greatly to its weight and durability and enhancing its value as a strong core in the tree trunk.

Vessels which are passing out of action often become partially or wholly filled by sac-like in growths from the living parenchyma cells which are in contact with them. The pits between parenchyma and vessels are not bordered onthe side of the former, so that they present a large surface of pit membrane, which is expanded by turgor pressure into the cavity of the vessel. Such ingrowths are called *tyloses*.They are ballon-like, cylindrical or even branched, and have thin walls, but occasionally they may be lignified, either like stone cells or like the walls of the vessel in which they lie. What determines these differences is unknown. The young vessel elements usually expand laterally very rapidly, reaching their fully diameter even while still in contact with the cambium, and creating considerable tissue pressures, while tracheids do not thus expand; yet there is no invariable distinction of size, though normally the expanded vessels are larger than the tracheids of the same wood.

While vessels are normally the channels of long distance transport, tracheids serve chiefly for abort distances and for water storage. Offshoots of the main bundles, including the leaf traces the leaf veins, consists almost entirely of tracheids. This distinction of function implies an advanced state of specialization and is characteristic of the Dicotyledons. Fibres play a large part in the make up of the secondary wood. Firstly there are the *fibre-tracheids*, thicker walled than the true tracheids but with the same bordered pits, and in some cases not sharply distinguishable. These grade through intermediate elements, increasingly narrow, thick-walled and elongated, to true *libriform fibres*. Libriform fibres in the wood are often as much as seven times as long as the cambial initials. Measurements show variations between 0.3 and 1.3 mm. Nevertheless this is small compared with the bast-fibres, which are seldom less than 1.0 mm. long and may reach 7.7 mm. in Urtica, or even 40 cm., i.e., 16 in., in Boehmeria nivea. In both cases, however, the elongation during differentiation is very considerable. As it takes place in regions where general growth in length has ceased, it has been held that the elongation of the individual cells must imply slipping or sliding of the cells on each other, and this has been called *sliding growth.*

The idea is open to some anatomical objections, but it cannot be said that the alternatives suggested are very convincing. One counter-argument, usually regarded as weighty, is that the pits in the walls of adjacent fibres show a normal degree of correspondence, which is considered to be impossible if the walls have slipped during growth. It is not impossible, however, that this correspondence may be secondarily established, since it is also found in the pits between adjacent tyloses in wood venica, in which cane the cells have manifestly only come into contact at a late stage of development.

The last class of fibres is that called *substitute fibres*, which are living cells, often containing starch, but with lignified walls. They are not clearly separated, except in point of length, from xylem parenchyma, and indeed they are sometimes divided by transverse walls forming a short chain of typical parenchyma. The xylem parenchyma cells arise by repeated transverse divisions of cambial initials. Their walls are lignified but not

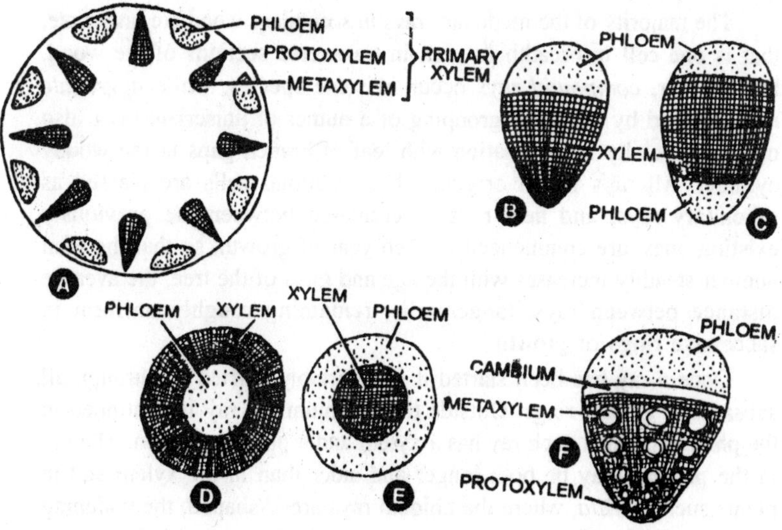

Fig. 1.18. Vascular bundles: A—radial; B—collateral open; C—collateral closed; D—bicollateral; E—concentric amphicribal; F—concentric amphivasal.

heavily, and they are richly supplied with simple pits on the radial and horizontal walls. They are important as stores of reserve substances, either starch, in most hard-wooded trees, or oil in soft-wooded trees. They also form important links between the cells of adjacent medullary rays.

Careful investigation shows that they are seldom isolated, but occur in irregular tracts, which contact one or more of the medullary rays, usually two, between which they form a tangential bridge. In this way they serve to make an intercommunicating system of living cells in wood, providing for tangential transport, as the medullary rays themselves do for radial transport. The distribution of the xylem parenchyma is highly characteristic of particular species of wood and is one of the features by which they may be identified. There are two main types:

1. *Apotracheal*, which includes those occuring in tracts which are independent of the vessels, i.e., metatracheal, and also the rarer cases where they are diffuse or *dispersed*.
2. *Paratracheal*, which includes all cases where they are grouped around the vessels.

In Conifers the xylem parenchyma usually occur only in the last row of elements in each annual ring. This arrangement, called *terminal*, is very rare in Angiosperms, being known only in *Magnolia* and *Salix*.

Medullary Rays

The majority of the medullary rays in secondary wood are uniseriate, that is one cell in breadth as seen in transverse sections of the wood, but broader, *compound rays* occur in some species, while *aggregate rays*, formed by the close grouping of a numer of uniseriate rays, also occur, especially in association with leaf of branch gaps in the woody cylinder. All rays which originate from cambial cells are classed as secondary rays, and new rays, intercalated between the previously existing ones, are commenced in each year of growth, so that the total number steadily increases with the age and girth of the tree, the average distance between rays, tangentially, remaining roughly uniform in successive rings of growth.

Once a ray has been started it is invariably carried on through all subsequent growth rings. Outside the cambium the ray is continued in the phloem, so that each ray has a xylem and a phloem portion. The ray in the phloem may be both longer and sider than in the xylem and in plants such as *Tilia*, where the phloem rays are V-shaped, the widening is due to serial cell divisions resembling cambial growth. Medullary ray initials, which are much shorter than other cambial cells, cut off cells both inwards and outwards by successive tangential divisions, though many more are added to the xylem rays than to the phloem rays, corresponding to the more rapid formation of xylem. The cells cut off are directly converted into parenchyma and added to the rays without further divisions.

The xylem ray cells become lignified like the other xylem parenchyma, the phloem ray cells are thin-walled and pitted, like phloem parenchyma. During the winter, when the cambium is dormant, the ray initials in the cambium take on the character of phloem ray cells, so that there is then no break in the radial continuity of the two portions of the ray. New initial cells are formed in spring by the tangential division of the inner-most phloem ray cell. Most rays are comparatively short in the vertical direction, some no more than one cell in height, and the tallest are not usually more than twenty cells.

The cells are, with few exceptions, uniform in character, and all contain living protoplasm, nuclei, and reserves of food material. They remain living without growth or division for as long as the plant exists, and the ray cells of heart wood in old trees are the oldest living cells nown. They may persist for an indefinite period without growth and without division. Medullary rays in woody Dicotyledons have been classified into a number of types on the following characters:

1. Whether both uniseriate and multiseriate rays are present, or only one kind.
2. Whether the cells composing the rays are all alike short, or whether some are vertically elongated.

The heterogeneous type, with both kinds of rays and both kinds of cells, is regarded as the most primitive, and the homogeneous type with only multiseriate rays of short cells is probably the most advanced. Where the xylem ray cells about on vessels, the ray field, i.e., the area of contact between ray cell and vessel, has simple, not bordered pits, which are sometimes single and very large, but more often small and either circular or elliptical.

Types of Wood

At the resumption of activity in spring the cambium produces comparatively large, thin-walled, rapidly expanding xylem cells, the *spring wood*, which meets the need for active conduction of water to the developing buds. At the height of summer cambial growth is reduced, but it again increases to a secondary maximum in autumn, associated with increased rainfall and an abundance of photosynthetic carbohydrates. The *autumn* wood is, however, much denser than the spring wood and contains a higher percentage of fibres. It is the dense zones of autumn wood which mark the "grain" of commercial timbers. Tracheids, vessels and fibres may be formed at any time during the year's growth, and the pattern of their distribution is often on index to the identity of the wood and decides its physical characteristics. Two alternative patterns affecting the vessels are so well marked that woods may be classified by them at a glance.

In one class the formation of vessels is almost or entirely confined to the spring, so that each annual ring begins with a zone of large vessels. This is called *ring-porous* wood. In the other class vessels are formed throughout the year. Such wood is called *diffuse-porous*. The vessels in these woods are nearly always smaller than in the ring-porous type, and are sometimes scarcely broader than the tracheids. They are also apparently much shorter. The total length of the open lumen in a single vessel is difficult to determine, but the passage of coal-gas through the wood gives us some idea of their relative lengths. Wood without vessels is completely impermeable to gas, but it will pass through a 10 ft. length of the ring-porous wood of Ash in sufficient amount to ignite, while more than 2 ft. of the diffuse-porous wood of the Maple stops it completely.

A small group of genera belonging to the order Ranales, namely *Drimys*, *Trochodendron* and *Tetracentron*, is notable for having no wood vessels at all. These woods are spoken of as homoxylous, for they consist of radially arranged tracheids of uniform width, recalling the wood of *Pinus*. The resemblance is carried further, for they have large bordered pits on the radial walls of the tracheids, like those of Conifers, though the pitting on the spring wood is scalariform. Whether this is significant from the evolutionary standpoint, as indicating a primitive character in these three genera, or whether they have lost their vessels by reduction in the course of evolution is a matter of opinion.

Each year's complement of secondary thickening forms one *annual ring*. The change from the dense, small-celled autumn wood to the soft, open type of spring wood is normally well-marked and unmistakable in all temperate trees, but in trees grown in an equatorial climate growth is continuous and the annual rings may be non-existent. The older fossil woods have either no annual rings or vague rings with no marked break between them, showing that there were no marked seasons even in temperate latitudes in Carboniferous time. A marked differentiation of climate first appears in the Mesozoic period. Annual rings are an excellent record of climatic variations, being wider or narrower according to the quality of the season and having a reduced average width in high latitudes and at high altitudes. It does not follow, as might be imagined, that narrow rings imply stronger wood. On the contrary, unfavourable conditions limit chiefly the development of the autumn wood, and narrow rings may consist mainly of the spongy spring wood. The width of the annual ring is not uniform all over the tree but is greater in the upper part, associated with a longer period of cambial activity.

Annual rings are usually symmetrical in vertical stems, but in oblique or horizontal branches, especially near the base, they become asymmetrical, the lower half of the upper half being thicker than the other and consisting of differently thickened elements. In Conifers the lower wood is reddish, the upper white, the former having much thicker cell walls. The difference is attributed to the mechanical effect of the weight of the branch, the upper wood being called *tension-wood* and the lower portion *compression-wood.* Vertical stems subject to consistent wind pressure from one side may show the same differences. Severe frost may cause upward or downward bending of large branches according to whether the thinner-walled tissue lies on the lower or the upper side of the branch, as is remarked in "Lorna Doone".

In the contracted stems of rosette plants successive rings of wood may not be formed at all, or if they are, they are related to environmental conditions, not to age. A constant relation to annual growth is only maintained in main stems of unlimited growth. In short branches, such as fruits spurs, annual rings are only formed in the first few years; while in thorns, unless they bear leaves, there is never more than one ring, however old they may be. The phloem rarely shows indications of annual rings. The periodicity of cambial growth is less marked on the phloem side. Activity begins later and may, in some cases, continue into and perhaps through the winter, the spring change in the phloem not being the formation of new elements, but the very marked swelling of the old elements through water intake.

Towards the end of its life the phloem may, though rarely, become signified, as may be seen in old stems of the annual Sunflower. Wood anatomy has proved a very useful adjunct to systematic Botany and is serving to supplement the ideas of classification and evolution based upon external morphology. The features principally used for comparison are: fibre types, distribution of the xylem parenchyma, type of perforation in the end walls of vessels and the presence of absence of "storied structure" in the wood. The origin of storied structure has been explained under the heading of Cambium. It implies a marked uniformity of length in each series of tracheids and vessel elements as seen in longitudinal section, so that the wood appears to be built up of vertically successive layers, or storeys. These characters are not always sharply marked and a good deal has to be left to the judgement of experience in deciding systematic affinities on wood characters. The above characters are grouped as specialized (i.e., advanced in the evolutionary sense) or unspecialized, as follows:

Unspecialized	*Specialized*
Scalariform vessel pores	Simple vessel pores
Fibre tracheids	Libriform fibres
Apotracheal parenchyma	Paratracheal parenchyma.
Diffuse-porous	Ring-porous
Unstoried structure	Storied structure

Specialized unspecialized features are widely distributed among angiospermic families and show no relation to specialized or unspecialized floral structure, but truly natural families are usually fairly homogeneous in their type of wood structure.

The Pith

The comparative anatomy of *pith* is still to be written, though it is evidently a very variable tissue. Many herbaceous plants, notably among the Umbeliferae and Compositae, have hollow stems, due to the failure of the pith to keep pace with the expansion of the outer tissues. The pith thus becomes cracked and riven, or *fistular* as it is called, only a few scanty fragments of tissue remaining to line the cavity. Certain woody plants, of which the Walnut (Junlans regia) is the best known, have discoid pith, dut to transverse cleavage, which, consequent upon growth in length of the young shoot, separates the pith into a series of discs, 2 to 3 mm. apart. The pith often becomes sclerotic, especially at the nodes, and nodal plates of highly thickened cells may remain even when the pith has disappeared in the internodes. The outer zones of the pith are frequently smaller celled, more sclerotic and more persistent than the central portion, forming a conspicuous boundary layer, called in German "Markrande", though an international name has not been adopted.

2

Water in Plant Cells

Membrane of Plant Cells

The membranes of plant cells are much more complex in their structure and properties than the nonliving membranes. With very few exceptions the protoplasm of each plant cell is enclosed by a more or less rigid wall which acts as a membrane. Through this wall, which usually consists of several layers, each with a complex organization of its own, must pass all substances moving either into or out of cell. In addition to the structure of the wall itself, the role of the plasmodesms must be considered in accounting for the movement of substances from one cell to another. Lining the cell wall of all mature plan cells is a layer of cytoplasm. The cytoplasm, or parts thereof, constitutes a second layer through which substances entering or leaving the vacuole of a cell must penetrate. Both theoretical considerations and experimentally observed facts support the view that the two limiting layers of cytoplasm possess different physiochemical properties from the intervening cytoplasm and may be regarded as distinct membranes.

The cytoplasmic membrane adjacent to the cell wall is called the plasmalemma; that enclosing the vacuole the tonoplast or vacuolar membrane. Since the boundaries between the cytoplasm and the cell wall and the cytoplasm and the vacuole constitute interfaces, certain protoplasmic ingredients probably become absorbed there in greater concentrations than in the interior of the cytoplasm. For this reason alone the postulation that the cytoplasmic membranes possess different properties from the interior cytoplasm seems justified. Results from several types of experimental investigations support this hypothesis. The immiscibility of protoplasm with water, already discussed, appears to

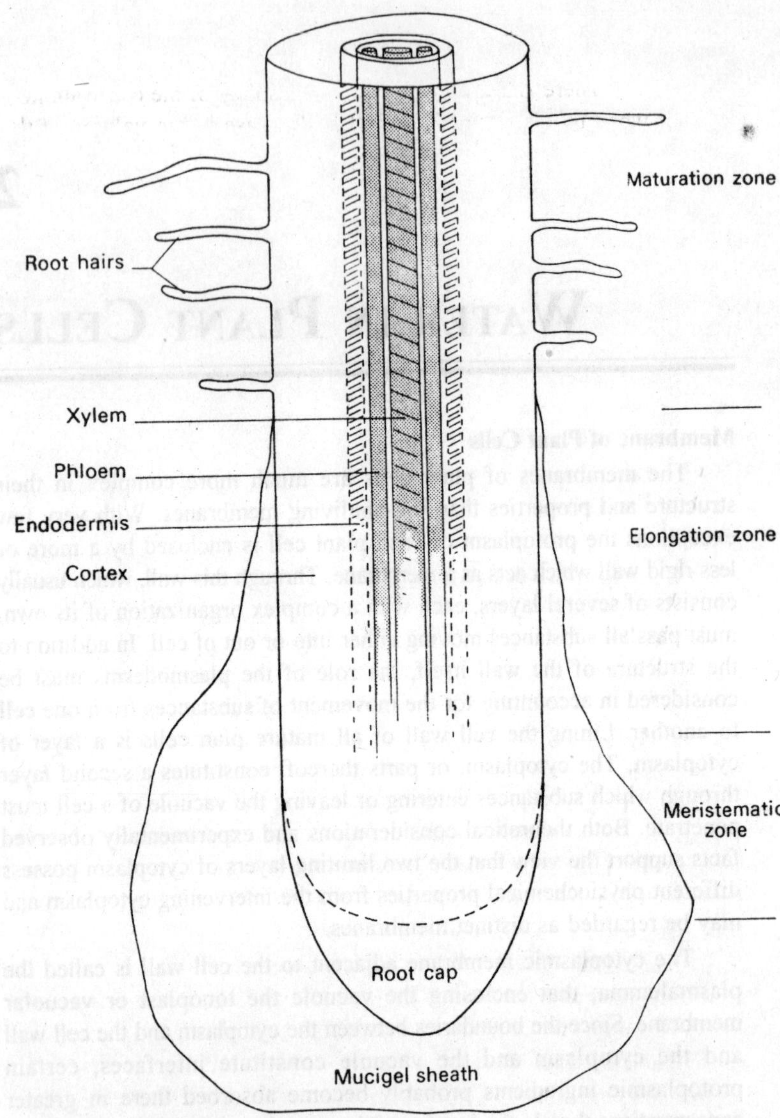

Fig. 2.1. Diagrammatic longitudinal section of apical zone of a plant root.

result at least partly from the fact that protoplasm is coated with a layer of material which is insoluble in water.

By means of micrcpipettes certain nontoxic dyes can be injected into the body of the cytoplasm through which they will spread rapidly. They do not, however, pass through either the plasmalemma or the tonoplast. Neither will they pass into the interior of the cytoplasm from a solution bathing the cell, nor from the vacuole if the latter is injected with the dye. This is direct evidence that the constitution of the plasmalemma and the tonoplast is different from that of the intervening cytoplasm. The presence of cytoplasmic membranes can also be demonstrated by the manipulation of living protoplasm with fine glass needles. It has been found possible to withdraw a plasmolyzed protoplast from its cell wall and to strip off the protoplasm from the vacuole leaving the latter as a free-floating sac full of cell sap enclosed by a delicate membrane—the tonoplast.

Most evidence indicates that the tonoplast and probably also the plasmalemma are thin liquid films which are immiscible with water. A complete picture of the permeability of a cell to different compounds can be drawn only in terms of the permeability of both the cell wall and the cytoplasmic membranes. In entering a cell a substance must first pass through the various layers of the cell wall, thence in turn through the plasmalemma, the interior cytoplasm, and the tonoplast before reaching the vacuole. Although some investigators prefer to refer all permeability phenomena of plant cells to the cytoplasm as a whole, the evidence that the plasmalemma and the tonoplast exist as distinct membranes with the property of differential permeability must be regarded as very substantial.

Permeability of Cell Membranes

In general, cell walls composed of cellulose and pectic compounds, such as those in most parenchymatous tissues, are quite permeable to water and solutes. Water and solutes probably penetrate such walls through the hydrophilic intermicellar material. Lignified walls are also quite permeable to water and solutes, but cell walls in which suberin or actin are present in appreciable quantities, on the other hand, are less permeable both to water and solutes. In contrast with the walls of most parenchymatous cells, the cytoplasmic membranes exhibit a high degree of differential permeability.

The term "cytoplasmic membrane" is here employed in a loose sense to refer to the entire cytoplasmic system of membranes : plasmalemma plus interior cytoplasm plus tonoplast. Actually many substances entering plant cells are intercepted and utilized it one way or another in the cytoplasm, and do not diffuse in appreciable quantities through the tonoplast. Similarly compounds synthesized in the cytoplasm may pass out of the cell without crossing the tonoplast. Hence the plasmalemma is often regarded as the most important unit in the cytoplasmic system of membranes. It should be realized, however, that substances can move from one cell to another, through the plasmodesms, without crossing the plasmalemma. Very few general statements can be made regarding the permeability of the cytoplasmic membranes to specific substances which will hold for all or even most plant cells. One such statement which can be made is that the cytoplasm is usually relatively permeable to water. This is also true for certain other small molecules, such as ammonia.

Permeability of the cytoplasmic membranes to most solutes of physiological importance such as sugars, electrolytes, and amino acids, on the other hand, appears to be very variable. Studies of many kinds of cells, for example, indicate that their cytoplasmic membranes are relatively impermeable to sucrose and other sugars. The slowness with which many kinds of plant cells recover from plasmolysis in sugar solutions is also evidence of the relative impermeability of the cytoplasm to such compounds. Nevertheless there is much indirect evidence that sugars often move into and out of many kinds of cells at relatively rapid rates. Similarly, although many permeability studies indicate the cytoplasmic membranes to be relatively impermeable to electrolytes, they often penetrate relatively rapidly into plant cells, when in any actively metabolizing condition.

The probable explanation of these apparent discrepancies is that most studies of permeability have been made upon mature cells which have lost most or all of their capacity for rapid absorption of electrolytes and sugars. The permeability of the cytoplasmic membranes to specific substances is therefore not a fixed property, but one which differs from one kind of cell to another and varies in the same cell with the age of the cell, the environmental conditions to which it is subjected, and other factors. Furthermore, there are many indications that the cytoplasm is not merely a passive membrane like a sheet of collodion, but that it, at times at least, participates actively in the movement of molecules and ions into or out of cells, or from one part of a cell to another.

Plasmolysis

If a vacuolate plant cell which is at least partially distended is immersed in a hypertonic solution of some substance to which the membranes are relatively impermeable, a characteristic series of changes takes place in the appearance of that cell. The first detectable occurrence is a gradual shrinkage in the volume of the entire cell as a result of outward osmosis. Consequently there is a reduction in the pressure exerted by the cell sap against the protoplasm and cell wall. This shrinkage in volume can be detected in many cells by measurement under a microscope, although there are some types of plant cells in which little change in volume occurs under such conditions.

There is a lower limit to the elasticity of the cell wall, however, and when this is attained no further decrease in the volume of the cell will occur. Since the cell wall is quite freely permeable to the water and the solutes of the surrounding solution, and hence the hypertonic solution is in contact with the outer surface of the protoplast, the cell sap will continue to lose water by outward osmosis just as if no cell wall were present. Hence the protoplasm will continue to shrink in volume after the contraction of the cell wall has ceased. At this stage, therefore, the protoplasmic layer will begin to recede from the cell wall. If the hypertonic solution is strong enough, this separation of the protoplasm from the cell wall will become very pronounced.

In some types of cells the protoplasm will "ball up" into a more or less spherical mass within the cell. More often the shrunken protoplasm assumes other configurations. This phenomenon is called *plasmolysis*. The pattern followed by the shrinkage of the protoplasm in plasmolysis is more or less typical for each kind of cell, although it may be modified somewhat depending upon physiochemical conditions within the protoplasm and the kind of solute used in the plasmolyzing solution. The space between the cell wall and the protoplasm will be filled, after the separation of the latter from the wall, with the external solution. If a plasmolyzed cell is immersed in water, it will slowly recover and regain a turgid state at a result of osmotic movement of water into the vacuole. Similarly, if immersed in a solution hypotonic to the cell, sap recovery will also ensue, but the degree of targidity attained will be less than if the cell were immersed in pure water. Ripid "deplasmolysis," however, results in the death of many kinds of plant cells.

Osmosis and Osmotic Pressure

Like gases and solutes, liquids exhibit diffusion phenomena. If, for example, water is brought into contact with another liquid such as ether, with which it is only slightly miscible, a slow diffusion of water molecules into the either will occur. Simultaneously a slow diffusion of the molecules of the either will take place into the water. Such diffusion will continue until the two liquids are mutually saturated.

Osmosis

This is by far the most familiar process involving the diffusion of liquids. An understanding of the dynamics of this process and of the significance of the physical quantity termed osmotic pressure is essential

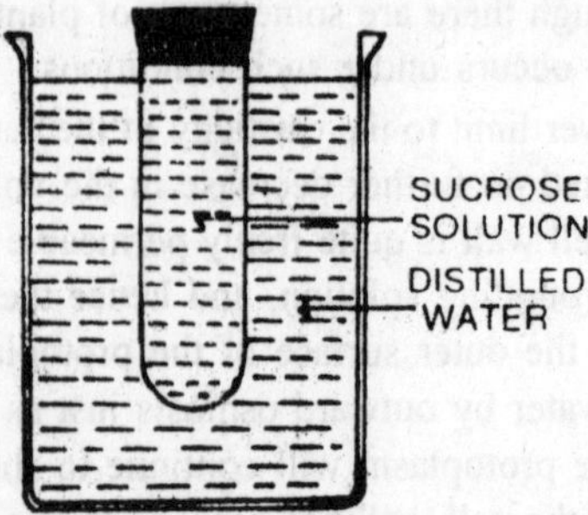

Fig. 2.2. Apparatus for demonstration of osmosis through a collodion membrane.

to an interpretation of the water relations of plant cells and tissues. A sac-like membrane of collodion is completely filled with a strong sucrose solution and immersed in a beaker of water. The top of the sac is tightly plugged with a rubber stopper. The collodion membrane is prepared so that it is permeable to water, but impermeable or practically so to sucrose; in other words, it is *differentially permeable*. It is also slightly elastic. After a short time the originally limp sac becomes rigidly distended. This is a result of the diffusion of water through the collodion membrane into the interior of the sac. Such a diffusion of water is an example of *osmosis*.

The pressure developed as a result of the entrance of water prevails throughout the solution and is also exerted against the inside wall of the membrane. If the membrane is virtually impermeable to the solute, eventually the entire system will come to equilibrium, after which there will be no further increase in the volume of water inside of the membrane. When the movement of the solvent is referred to in discussing osmosis it is always the net movement which is meant. Solvent molecules will always be moving across the membrane in both directions but, except when an equilibrium has been attained, more molecules will move per unit of time in one direction than in the other. As will be shown in the later discussion this net movement is always from the region of the greater diffusion pressure of the solvent molecules to the region of their lesser diffusion pressure.

The maintenance of the diffusion pressure gradient of the liquid in most osmotic phenomena is made possibly by the relative impermeability of the membrane to the molecules of solutes. If the membrane were permeable to the solute particles, they would diffuse outward across the membrane, while the molecules of the solvent were diffusing in the opposite direction; this would be another example of the principle of independent diffusion. Osmosis, the diffusion of a *solvent* (not a solution) across a differentially permeable membrane, may be regarded as a special case of diffusion.

Osmosis occurs whenever two solutions with a common solvent, in which the diffusion pressures of the solvent are different, are separated by a membrane more permeable to the solvent than to the solutes. In living organisms water is the only important solvent which moves by osmosis, hence the following discussion of this process will be in terms of water and aqueous solutions. Diffusion of the molecules or ions of a solute through a membrane is also common phenomenon but occurs independently of the diffusion of the solvent molecules, the accordance with differences in the diffusion pressure of the solute, and seldom in amounts which are proportional to the diffusion of the solvent. The term "osmosis" is sometimes applied to such a diffusion of a solute across a membrane; likewise it is sometimes applied to the diffusion of a gas across a membrane. Historical considerations, however, give priority to osmosis as a term for the diffusion of a solvent across a membrane. Furthermore, certain aspects of this phenomenon are so distinctive that it is convenient to have a distinguishing term for it.

Membranes and Permeability

The distinctive aspects of osmosis as compared with other diffusion processes result largely from the presence of the differentially permeable membrane. The concept of permeability is inseparable from the idea of a membrane. Permeability is a property of the membrane, not of the substance which diffuses through it. Thin layers or sheets of many different kinds of substances, such as rubber, parchment paper, collodion, cellophane, gelatin, and copper ferrocyanide, may serve as membranes. Some membranes are impermeable to all, or virtually all, substances; others allow all or most substances to diffuse through them with little impediment.

Many biologically important membranes are of the type known as differentially permeable. Such membranes allow some substances to pass through them much more readily than others. They may be, and often are, impermeable or virtually so to some substances, while others diffuse through them quite freely. Alternative but less desirable terms for *differentially permeable* are *semi permeable* and *selectively permeable*. The simplest kinds of membranes are essentially molecular sieves. The pores of a given membrane of this type are such that micelles, molecules and ions below a certain size can diffuse through them, while those exceeding this size cannot pass through. Membranes of collodion, parchment paper, and copper ferrocyanide owe their differential permeability largely to their sieve-like properties. Some membranes are differentially permeable because certain substances are more soluble in them than others.

The rubber balloon experiment described furnishes an example of such a membrane. Carbon dioxide gas is much more soluble in rubber than oxygen or nitrogen, hence a rubber membrane is much more permeable to carbon dioxide gas than to either of the other two. Another example of a membrane of this type in operations can be seen in an experiment set up.

A thin layer of water is introduced on top of a layer of chloroform in a test tube, the tube nearly filled with either, and stoppered. A second tube is prepared in the same way except that xylene is substituted for the ether. After several days it will be observed that the level of the water layer in the first tube has risen, while in the second tube it has fallen, although the distance through which the layer moves is not as great in the second tube. In the first tube either is osmosing through the water *membrane* more rapidly than the chloroform; hence the volume of liquid below the membrane is increasing and the layer of water rises. In the second tube chloroform osmoses through the water membrane more rapidly than the xylene; hence the water layer falls in the tube. Much less chloroform osmoses in a given time through the water layer in the second tube than does either in the first tube. Of these three compounds, either is the most soluble in water, chloroform next most soluble, and xylene least soluble. Permeability of the water membrane to these three compounds is clearly correlated with their solubility in water.

Diffusion Pressure of Liquids

Like a gas, water or any other liquid may be considered to possess a diffusion pressure, although the existence of such a physical quantity in liquids becomes apparent only under conditions as, for example, when a solvent and a solution are separated by a differentially permeable membrane. Diffusion pressure is the cause of diffusion not its result. Just as gases have a pressure, whether or not any diffusion is occurring, so do liquids and solutes have a diffusion pressure, whether or not any diffusion is occurring. The only two factors which influence the diffusion pressure of pure water are pressure and temperature.

If pressure is imposed on water, as by a piston in a closed system, or by confining effects of the walls of a closed osmometer or plant cell when the volume of the enclosed water is expanding as a result of osmiosis, the diffusion pressure of the confined water increases by the amount of the imposed pressure. If the imposed pressure is 8 atm., for example, the diffusion pressure of the water increases by 8 atm. Correspondingly, if water is subjected to a "negative pressure" (tension), its diffusion pressure decreases by the amount of the negative pressure. The effect of the imposed pressure on the diffusion pressure of the liquid is quantitatively the same whether the liquid is pure, or whether solutes are present. Although the diffusion pressure of a liquid is also theoretically affected by temperature, it is usually not necessary to consider this effect in evaluating the mechanism of the diffusion of water in plants. In an aqueous solution, however, the diffusion pressure of the water is influenced by still another factor the proportion of solute practices (ions plus molecules) to solvent molecules. The greater this proportion, the less the diffusion pressure of the water.

Within a wide range of concentrations of the solute molecules or ions, the diminution in diffusion pressure is closely proportional to the ratio of solute particles to solvent molecules. Ideally all diffusion phenomena including osmosis, should be interpreted on the basis of differences in diffusion pressures. Diffusion phenomena in gases can be readily analyzed in this way because gas pressures can usually be ascertained. The exact magnitudes of the diffusion pressures of liquids, on the other hand, cannot readily be measured. Hence use of this physical concept in the quantitative interpretation of the diffusion of liquids is not feasible.

It is usually possible, however, to measure quantitatively the amount by which the diffusion pressure of the water in a given solution is less than that of pure water at the same temperature and under atmospheric pressure. This quantity is called the *diffusion-pressure deficit* or *pressure deficit* and can be used in interpretation of osmosis and other diffusion phenomena in which water molecules participate, even if we do not know the exact magnitude of the diffusion pressure of water. In this indirect manner it is possible to use the concept of diffusion pressure in the analysis of diffusion phenomena in liquids.

Osmotic Pressure

This term was originally employed to designate the maximum pressure which develops in a solution enclosed within an osmometer under certain ideal conditions. An osmometer is an apparatus for measuring the magnitude of osmotic pressures. If a manometer or pressure gauge were inserted through the stopper of an apparatus set up, the arrangement would serve as a crude osmometer. Most precise measurement of osmotic pressures have been made with osmometers constructed cylindrically-shaped porous clay cups in the pores of which have been precipitated differentially permeable membrane of cupric ferrocyanide. The pressure developed a measured with a sensitive mercury manometer or in other ways.

The necessary ideal conditions, only attained under rigorous experimental conditions, are that the membrane be permeable only to the solvent, that it be immersed in the pure solvent, and that the pressure equilibrium be attained without any appreciable dilution of the enclosed solution. The temperature of the experimental setup must also be controlled and recorded, as the osmotic pressure of a solution is in part a function of its temperature. The term "osmotic pressure" can be more usefully employed as an index certain properties of a solution, rather than

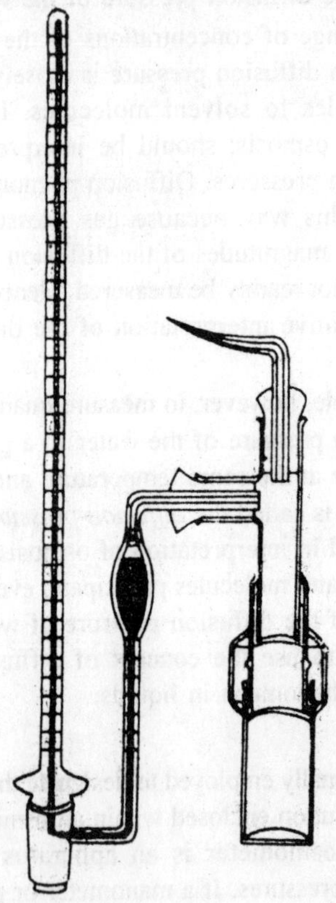

Fig. 2.3. Pfeffer's osmometer with attached closed manometer for measuring the pressure developed. The lower part of the osmometer is a porous clay cup in the walls of which have been precipitated a differentially permeable membrane of copper ferrocyanide.

as a designation for an actual pressure which is attained only under certain seldom realized conditions.

In part such a usage is a derivative one from the original usage described in the preceding paragraph. For example, a modal solution of sucrose standing in a beaker may be designated as having an osmotic pressure of 27 atm. At 25°C. Obviously this solution is not exerting an actual pressure of 27 atm., and this statement merely implies that it has certain properties and potentialities which will become apparent under particular circumstances. Such a statement regarding a solution is analogous to the statement that an idle electric motor has a rating of — horsepower; both statement are indices of potential rather than existing capacities. Used thus as an index, the osmotic pressure characterizes the solution in two ways.

In the first place, it tells us that the maximum possible pressure which could be developed in that solution if it were permitted to come to equilibrium with pure water in osmotic system at 25°C is 27 atm. The osmotic pressure is thus a rating of the *potential* maximum pressure which can be developed in the solution as a result of osmosis. In the second place, osmotic pressure is an index of the *diffusion-pressure deficit* of the water in a solution insofar as this results from the presence of solutes. Commonly used synonyms for osmotic pressure in this index sense are *osmotic concentration*, *osmotic value*, *osmotic potential*, and *osmotic power*.

Turgor Pressure

Turgor pressure is the actual pressure which develops in a closed osmometer or plant cell as a result of osmosis or imbibition (see later). Use of the term "osmotic pressure" in this sense should be avoided as it only leads to confusion. A given solution has a unique osmotic pressure at a given temperature, but its turgor pressure is variable. Ordinarily the turgor pressure of a solution lies in the range between zero and its osmotic pressure, but under certain conditions it may exceed the osmotic pressure, and under certain other condition it may be negative in value. The turgor pressure prevailing within a solution is also exerted against the confining walls of the system.

Quantitative Aspects of Osmosis

Let us suppose that a solution with an osmotic pressure of 20 atm. Completely fills an essentially inelastic membrane which is permeable only to water and is arranged. Let us further suppose this membrane to be immersed in pure water, which has, of course, a zero osmotic pressure. Both the solution within, and the water without, the membrane are initially under atmospheric pressure only, the effect of which (equivalent to a turgor pressure of 1 atm.) is conventionally disregarded when all parts of the system are under atmospheric pressure. From the previous discussion it should be clear that the diffusion-pressure deficit of the water in the enclosed solution is 20 atm that of the water outside of the sac, zero. Water, therefore, osmoses across the membrane into the solution. The entering water engenders a gradually increasing turgor pressure throughout the solution in the sac. Subjection of the water in this solution to such a pressure raises its diffusion pressure and hence lower its diffusion-pressure deficit. An equivalent increase takes place in the diffusion pressure of the solutes present.

A dynamic equilibrium will be attained when the turgor pressure of the internal solution reaches 20 atm. At this point the 20 atm. diffusion-pressure deficit which would otherwise result from the presence of solutes is offset by the 20 atm. Turgor pressure and the diffusion-pressure deficit of the water in the solution is zero, which is the same as that of the pure water outside the membrane. When the diffusion-pressure deficit of the water on both sides of the membrane becomes equal, osmosis ceases, and the number of water molecules moving through the membrane in the inward direction will be the same as the number moving in the outward direction per unit of time. Since the membrane is assumed to be inelastic or nearly so, osmosis of only a very small amount of water into the solution is necessary to raise its turgor pressure to 20 atm.

Since there is virtually no dilution of the solution, its final osmosis pressure, for all practical purposes, is the same as its initial osmotic pressure. In living organisms water usually osmoses across a membrane from one solution to another rather than from pure water into a solution. Let us consider, therefore, an analogous situation, exactly like that described above, except that the water on the outside of the membrane is replaced with a solution of 12 atm., osmotic pressure. Initially, before any turgor pressure develops, the diffusion-pressure deficit of the internal solution in this system is 20 atm.; that of the external solution, 12 atm. Water therefore osmoses across the membrane into the solution within the sac.

As the turgor pressure of the internal solution increases, the diffusion-pressure deficit of the water in that solution decreases. When a turgor pressure of 8 atm. has been attained, the diffusion-pressure deficit will have decreased to 12 atm. which is the same as that of the external solution. Dynamic equilibrium is thus attained in this particular system when the turgor pressure prevailing in the internal solution reaches 8 atm. In the foregoing discussion an inelastic membrane has been assumed.

Many membranes, however, especially in living organisms, are elastic within limits. When the membrane is appreciably elastic, the further complication of shifts in the osmotic pressure of the internal solution during osmosis is introduced. Let us assume an osmotic system identical with that described in the preceding paragraph in the preceding paragraph, except that the membrane is elastic enough to permit a 25 per cent increase in the volume of the internal solution before a dynamic equilibrium is attained. As the internal solution increases in volume it becomes more dilute, i.e., less concentrated. The osmotic pressure of a solution decreases with a diminution in its concentration. If we assume a proportionate decrease in osmotic pressure with decrease in concentration, which would be strictly true only in certain kinds of solutions, the final osmotic pressure in the internal solution would be 16 atm. (20: X=125:100).

A turgor pressure of only 4 atm. would be required under these conditions to bring the water in the internal solution into dynamic equilibrium with that in the external solution which has an osmotic pressure of 12 atm. In such a system the turgor pressure of the internal solution is increasing while its osmotic pressure is decreasing. The only difference in end result when the membrane is elastic is that the turgor pressure and osmotic pressure at dynamic equilibrium are somewhat less than they would be otherwise. If the volume of the external solution is large relative to that of the internal solution, variations in the osmotic pressure of the former are usually negligible. It should be evident from the foregoing discussion that the diffusion-pressure deficit of the water in a solution is always equal to the osmotic pressure of the solution less the turgor pressure which prevails in it. This relation can be conveniently expressed in the simple equation.:

$$\text{D.P.D.} = \text{O.P.} - \text{T.P.}$$

As applied to the initial diffusion-pressure deficits in the second example given above these equation would work out as follows:

External solution : D.P.D. = 12 – 0 = 12 atm.

Internal solution : D.P.D. = 20 – 0 = 20 atm.

At dynamic equilibrium the diffusion-pressure deficit of this same external solution is unchanged; that of the same internal solution is then calculated as follows:

$$\text{D. P. D.} = 20 - 8 = 12 \text{ atm.}$$

Magnitude of the Osmotic Pressures in Plant Cells

In many discussions of the osmotic pressure of plant cells no distinction is made between the "osmotic pressure of the cells at incipient plasmolysis" and their osmotic pressure under more or less turgid conditions. When the term osmotic pressure is employed without qualification it can be assumed to refer to the osmotic pressure of cells in their usual distended condition. It is generally considered that the results of determinations by the cryoscopic method represent approximately the osmotic pressure of the cells in tnis condition. Different organs or tissues of the same plant may exhibit a wide range of osmotic pressures. Even similar organs on the same plant—leaves, for example—may vary considerably among themselves in the average osmotic pressure of their cells.

Furthermore the tissues within the same organ usually show a considerable variation in osmotic pressures. The tatsophyll cells of most leaves, for example, show higher values than the epidermal cells. In general, the osmotic pressures of the plant cells and tissues of the mesic species of North America vary in their extreme range from a fraction of an atmosphere to about 50 atm. Most of the values lie, of course, within a narrower part of this range, probably within the limits of a 4 to 30 atm. span of values. The osmotic pressure of a given plant cell or tissue is by no means a fixed quantity, but often fluctuates considerably. More or less regular daily or seasonal variations occur in the osmotic pressures of many kinds of plant cells or tissues.

Factors Influencing the Osmotic Pressure of Plant Cells

Any factor which influences either the water content of a plant cell or the solute content of its cell sap will have an effect on the magnitude of the osmotic pressure of that plan cell. The water content of the plant as a whole, and hence of its constituent cells, is controlled principally by the relative rates of transpiration and the absorption of water. The later process is influenced very markedly by the water content and other conditions prevailing in the soil. Individuals of the same species invariably have a higher osmotic pressure when growing under drought conditions than when provided with a favourable water supply. This is at lease partially a result of the relatively lower water content of the leaves which obtain when the available soil water supply becomes low. Other factors which are also probably involved are a decrease in the growth rate of the plant which often permits an accumulation of mineral salts and soluble foods, and a shift of the starch ↔ soluble carbohydrates equilibrium toward the side of the soluble carbohydrates.

Table 2.1. The effects of different soil water contents upon the osmotic pressure of maize plants

Water contents of Soil, Per cent dry weight	*Osmotic pressure of tops*	*Osmotic pressure of roots*
31 per cent	22.06 atmos	5.9 atmos
25	23.08	7.23
16	24.36	7.79
14	25.04	9.24
13	25.47	11.34
11	26.48	11.98

The solute content of the cell sap is controlled by the specific metabolic processes of the plant and by the absorption of mineral salts by the plant from its environment. The rate of photosynthesis is an important factor in determining the osmotic pressure of plant cells, particularly those of leaf tissues. The influence of the inherent metabolic processes of any species upon the kinds and concentrations of various types of soluble organic compounds produced, such as simple carbohydrates, organic acids, and amino acids has an important effect on the magnitude of the osmotic pressure in any species. Metabolic conditions and their effects upon osmotic pressures may also be altered by environmental conditions.

Table 2.2 The effect of the osmotic pressure of the soil solution upon the osmotic pressure of the roots of maize.

Osmotic pressure of soil solution	*Osmotic pressure of sap from root cells*
1.21 atm	4.59
1.99	5.48
3 38	6.61
4.96	7.51
7.22	8.19

A well-known example of this is the difference in the osmotic pressures of sun and shade leaves on the same plant. The former almost invariably has the higher osmotic pressure, presumably at least in parts as a result of their greater photosynthetic activity. The mineral salts which contribute to the osmotic pressures of plant cells are all absorbed from the soil or in aquatic species from the water in which part or all of the plant is immersed. Different species of plants vary greatly in their toleration of high concentration of mineral salts in the soil. All species can become adjusted, within limits, to a change in the mineral-salt content of the substratum. This adjustment takes the form of an increase in the osmotic pressure of the plant with an increase in the osmotic pressure of the medium from which it obtains its mineral salts.

Other tissues of plans also show an increase in osmotic pressure with increase in the osmotic pressure of the soil solution, but such increases are generally most pronounced in the roots. Species indigenous to saline soils usually have relatively high osmotic pressures. Such soils are rich in soluble salts, and the high osmotic pressures found in species native to such soils result from the absorption of relatively large quantities of mineral salts. In fact, the highest recorded osmotic pressure for any species of plant—202.5 atm.—was found in the saltbush (*Atriplex confertifolla*), a saline soil species. Salt marsh plants also generally have relatively high osmotic pressures. This is also a correlation with the relatively high salt content of the substratum. Except in halophytes, however, a larger proportion of the osmotic pressure of most kinds of plant cells results from the presence of organic solutes than of mineral salts. Root cells are more likely to prove exceptions to this last statement than cells in the aerial organs.

The Osmotic Quantities of Plant Cells

Let us suppose three similar plant cells, each in a state of incipient plasmolysis (i.e., with a zero turgor pressure), and each with a cell sap osmotic pressures of 10 atm. to be immersed, the first in pure water, the second in a solution with an osmotic pressure of 6 atm., and the third in a solution with an osmotic pressure of 10 atm. The volume of the liquid in which each cell is immersed is assumed to be very large in proportion to the volume of cells. In order to simplify the discussion, certain assumption, certain assumptions will be made regarding these cells and, unless stated to the contrary, regarding all other cells discussed in the remainder of this chapter. These assumptions are : (1) the cells are at the same temperature, (2) the cytoplasmic membranes are permeable only to water while the cell walls are permeable to both water and solutes, (3) the walls of all the cells are equally elastic, and (4) volume-changes in the cells are so small that any resulting effects upon the osmotic pressure of the cell sap are negligible and can be disregarded.

A review of the discussion of the quantitative aspects of osmosis in physical systems will be helpful in following the ensuing closely analogous discussion with respect to cells. Osmosis of water into cell A begins immediately upon its immersion, because the diffusion-pressure deficit of the water in the cell sap is 10 atm. and that of the surrounding water is zero. The entrance of water into the cell results in a gradually increasing *turgor pressure* within the cell, just as it does in an osmometer. This pressure not only prevails within the water mass, but is also exerted against the protoplasmic layer and, in turn, against the cell wall. The increase in turgor pressure therefore results in a gradual distention of the cell.

Imposition of a pressure of 10 atm. on the water in the cell sap will increase its diffusion pressure by 10 atm. Since the initial diffusion-pressure deficit of the water in the cell sap was 10 atm., imposition of a 10-atm. turgor pressure reduces its diffusion-pressure deficit to zero. This is also the diffusion-pressure deficit of the surrounding water, hence a dynamic equilibrium is attained when the turgor pressure in the cell sap reaches 10 atm. The diffusion-pressure deficit of the water in the cell sap of cell B is 10 atm., that of the water in the surrounding solution 6 atm. Hence there will be a net inward movement of water into the cell. The turgor pressure thus develops within the cell, but at its maximum under these conditions it will attain a value of only 4 atm. Since the initial diffusion-pressure deficit of the water in the cell sap was 10 atm., imposition of a turgor pressure of 4 atm. reduces this to atm. which is the diffusion-pressure deficit of the water in the surrounding solution.

Hence the attainment of a turgor pressure of only 4 atm. results under these conditions in an osmotic equilibrium. Since the magnitude of the turgor pressure developed in cell will be less than in cell A, the latter cell will be more distended than the former. Cell C will remain, according to the conditions prescribed for this experiment, in a state of incipient plasmolysis. The diffusion-pressure deficit of the water in the cell sap and in the solution are equal (i.e., their osmotic pressure are equal and neither is under any pressure, except, of course, atmospheric pressures); hence a dynamic equilibrium will be established as soon as the cell is immersed in the solution.

Since there is no net movement of water into the cell no turgor pressure will develop, a condition which will obtain in all plant cells in the condition of incipient plasmolysis. In cells A and B a dynamic equilibrium has been attained by an adjustment of the diffusion-pressure deficit of the water in the cell sap until it is equal to that of the water in the surrounding liquid. This adjustment was attained in each of these cells by a shift in the magnitude of the turgor pressure, which is one of the two components determining the diffusion-pressure deficit of the water in the cell sap, the other being the osmotic pressure. In cell C no shift in the magnitude of the turgor pressure occurred, because the diffusion-pressure deficit was initially the same in the cell sap and in the solution.

Unlike a solution exposed to the atmosphere, the diffusion-pressure deficit of a cell is not equal to its osmotic pressure except when the turgor pressure of the cell is zero. The term "diffusion-pressure deficit of a cell" should be considered as an abbreviation for "diffusion-pressure deficit of the water in the cell sap." The interrelationships among the osmotic pressure, turgor pressure, and diffusion-pressure deficit of a plant cell should be further clarified by a study. This diagram also takes into account the influence of volume changes in the cell upon these physical quantities. In the interests of simplicity the effects of such volume changes upon the osmotic quantities of the cell have been disregarded in the discussion up to this point. When the cell is completely flaccid (relative volume =1.0), its diffusion-pressure deficit is equal to its osmotic pressure (20 atm.) while its turgor pressure is zero. As the volume of the cell increases as a result of influx if water, the osmotic pressure decreases because of the dilution of the cell sap. The pattern of the change in magnitude of the turgor pressure with increase in cell volume is not the same in all kinds of cells, but often follows a trend.

The diffusion-pressure deficit, being equal to the difference between the osmotic and turgor pressures, shows a progressive decrease as the turgidity of the cell increases. When the cell attains a condition of maximum turgidity (relative volume=1.5), its turgor pressure is equal to its osmotic pressure while its diffusion-pressure deficit has fallen to zero. The dotted extension of the line for turgor pressure to the left indicates that it sometimes has a negative value. Conditions under which this may occur are discussed shortly. The diffusion-pressure deficit of a cell in which the turgor pressure is negative is greater than its osmotic pressure. In many types of cells volume changes are much less marked than indicated in this figure, and hence they can often be disregarded in generalized considerations of the water relations of plant cells.

In some species many of the cell have walls which are virtually inelastic. This is especially true of the tissues of xerophytes and some water plants. The fundamental interrelation between the osmotic pressure, turgor pressure, and diffusion-pressure deficit of a plant cell is the same as for an osmometer and is expressed by the same equation:

D. P. D. = O. P. – T. P.

The quantity to which the name "diffusion pressure deficit" has been given in this discussion has been variously termed by writers in the field of plant physiology. The more commonly used synonyms for this term are "pressure deficit," "turgor deficit," "net osmotic pressure," "suction force' and "suction tension." The osmotic pressure, diffusion-pressure deficit, and turgor pressure are collectively called the *osmotic quantities* of plant cells. As previous discussion has shown, a full evaluation of these quantities also requires a consideration of the volume changes of plant cells. In plant tissues many of the cells are under a pressure imposed upon them by the surrounding cells.

In addition to its own turgor pressure the water in such a cell is also subjected to this pressure of external origin. This pressure is just as much a factor in determining the diffusion-pressure deficit of a cell as its own turgor pressure. The true diffusion-pressure deficit of such a cell is therefore less than that of the cell considered as an individual unit by the amount of this added pressure. Hence the equation representing the factors determining the diffusion-pressure deficit of a cell under such conditions becomes:

D.P.D. = O.P. —T.P. — Pressure exerted by surrounding cells

The water in the vessels and cells of plants frequently passes into a state of tension. Imposition of a pressure from an external source raises the diffusion pressure of water. Imposition of a tension, which is in effect a negative pressure, has precisely the opposite effect. In cells and vessels this can happen only if the enclosed water shrinks in volume to such a

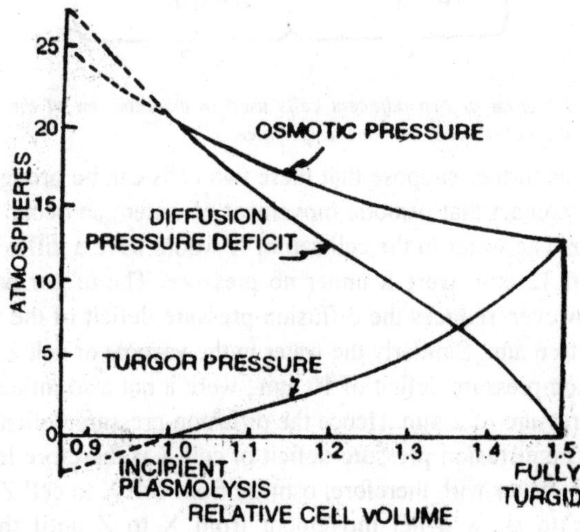

Fig. 2.4. Interrelationships among the osmotic pressure, turgor pressure diffusion-pressure deficit, and cell volume of a plant cell.

point that the encompassing protoplasm and cell walls are pulled inward as a result of adhesion between the walls and water. The resulting counter pull exerted by the walls on the water results in throwing it into a state of tension. Under such conditions the turgor pressure is negative in value and the tension (negative pressure) developed within the water will be equal to the negative turgor pressure. The equation for the diffusion-pressure deficit of a cell under such conditions becomes:

$$\text{D.P.D.} = \text{O.P.} - (-\text{T.P.})$$

In other words, the diffusion-pressure deficit of the water in a cell or vessel when subjected to a tension is equal to the osmotic pressure plus the tension (negative pressure) imposed on the water.

Dynamics of the Osmotic Movement of Water in Plants

In order to simplify this part of the discussion, changes in the osmotic pressure of the cell sap resulting from volume changes in the cells will again be disregarded, as usually these are not great enough to modify seriously any generalized picture of the water relations of plant cells. Let us imagine a certain cell (X) to have an osmotic pressure of 12 atm., and a turgor pressure of 6 atm., and a second cell (Z) to have an osmotic pressure of 10 atm. and a turgor pressure of 2 atm.

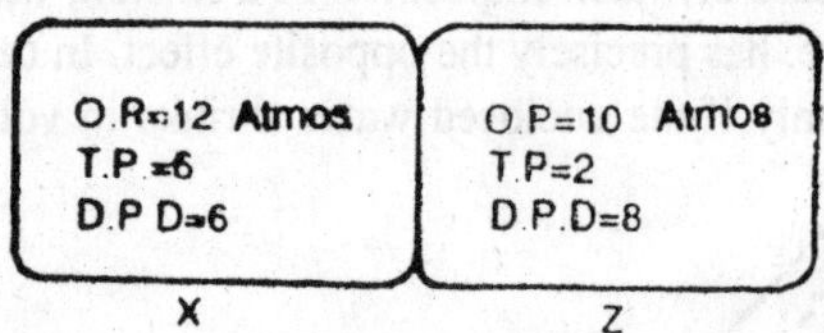

Fig. 2.5. Diagram of two adjacent cells used in explaination of the mechanism of the cell-to-cell movement of water.

Let us further suppose that these two cells can be brought into such intimate contact that osmotic movement of water can occur from one to the other. The water in the cell sap of X would have a diffusion-pressure deficit of 12 atm. were it under no pressure. The turgor pressure of 6 atm., however, reduces the diffusion-pressure deficit of the water in the cell sap to 6 atm. Similarly the water in the vacuole of cell Z would have a diffusion-pressure deficit of 10 atm., were it not also influenced by the turgor pressure of 2 atm. Hence the diffusion-pressure deficit of cell Z is 8 atm. The diffusion-pressure deficit of cell X is therefore less than that of cell Z. Water will, therefore, osmose from cell X to cell Z. Water will continue to show a net movement from X to Z until the diffusion pressure deficits of the two cells are equal.

In the movement of water from cell to cell in plants, it is the diffusion-pressure deficits and not the osmotic pressures which tend to equilibrate. This is only a special aspect of the fundamental tendency of the diffusion-pressure of water to attain a uniform value throughout any system. It is by no means impossible, therefore, for water to move from a cell of higher to one of lower osmotic pressure. After a dynamic equilibrium has been established between two cells their diffusion-pressure deficits will seldom be an exact average of their initial diffusion-pressure deficits. The only general statement which can be made is that, at equilibrium, the diffusion-pressure deficits of the two cells will be equal, and this value will lie somewhere between the two original values of the cells. Whenever the diffusion-pressure deficits of two adjacent cells are dissimilar, a diffusion-pressure-deficit gradient exists between them. Movement of water from one cell to another can occur only when such a gradient exists. Other conditions being equal, the "steeper" this gradient, i.e., the greater the difference in diffusion-pressure deficits, the more rapidly one cell will gain water from the other.

The term "diffusion-pressure deficit gradient" can also be applied to a chain of cells in which the diffusion-pressure deficit increase serially from cell to cell. Several examples of such gradients are discussed in subsequent chapters. The diffusion-pressure of water is influenced by temperature, although for small temperature difference the magnitude of this effect is not great. If two adjacent plant cells were to be maintained at different temperatures, theoretically this would have an effect on the rate or direction of osmosis. Actually, however, under the thermal conditions ordinarily prevailing in plants, this effect is so slight that it can be disregarded.

Methods of Measuring the Diffusion-pressure Deficit of Plant Cells

Most methods of measuring this quantity are based on the principle, previously discussed, that if a cell is immersed in a solution with an osmotic pressure equal to the diffusion-pressure deficit of the cell, a dynamic equilibrium is immediately established, and no change will occur in the volume of the cell because no net movement of water takes place. In solutions with a higher osmotic pressure than the diffusion-pressure deficit of the cell it will decrease in volume; in solutions of lower osmotic pressure than its diffusion-pressure deficit it will increase in volume. The "simplified method" of measuring diffusion-pressure deficits, introduced by Ursprung (1923), has been one of the most widely used.

In this method narrow strips of tissues are cut from such structures as thin leaves or petals. The length of each strip is immediately measured under the microscope, usually while mounted in paraffin oil. Several strips are then immersed in each of a graded series of sucrose solutions in which they are allowed to remain until an equilibrium has been attained in each solution, after which their lengths are measured. The osmotic pressure of the solution in which no change in the length of the strips occurs is considered to equal the average diffusion-pressure deficit of the cells in the strip. For the measurement of the average diffusion pressure deficit of the cells in bulky tissues such as potato tubers, masses of tissue such as a cylinders of approximately equal dimensions can be employed. The equilibrium point can be determined by measuring changes either in the weight or the volume of the cylinder.

The solution in which the cylinder neither gains nor loses weight for volume) is considered to have an osmotic pressure equal to the average diffusion-pressure deficit of the cells in the cylinder. Determinations of diffusion-pressure deficits by the methods described above should not be confused with plasmolytic determinations of the osmotic pressures of plant cells. In the latter type of determination the critical measurement is the osmotic pressure of the solution with which the cells come to an equilibrium without any turgor pressure, i.e., at incipient plasmolysis.

In diffusion-pressure-deficit determinations the critical measurement is the osmotic pressure of the solution with which the cells come to equilibrium without any change in their turgor pressure, i.e., without any change in the volume of the cells. Only when the cell is initially in a completely flaccid condition will determinations of its osmotic pressure and its diffusion-pressure deficit result in the same values. All of the methods of measuring the diffusion-pressure deficits of plant cells are subject to most of the flaws which are also inherent in plasmolytic determination ot osmotic pressure, as well as some other serious errors.

Except possibly for a few types of cells which are especially well adapted to such measurements, it is doubtful if the values obtained in determinations of diffusion-pressure deficits are ever more than fair approximations of the true values which obtain in plant cells. The turgor pressure of a cell cannot be determined directly. However, if the osmotic pressure of a cell or group of cells has been determined, and likewise the diffusion-pressure deficit, the turgor pressure can be calculated from the equations given earlier in the chapter.

The Imbibitional Mechanisms of the Movement of Water

In the preceding discussion, plant cells have been considered purely as osmotic systems. There is no doubt that the osmotic mechanism of the cell-to-cell translocation of water exists and operates in most if not all kinds of plant cells. Certain of the movements of water which occur in plants, however, result from the operation of an imbibitional rather than an osmotic mechanism. Whenever the diffusion-pressure deficit of water is greater in an imbibant than in a contiguous solution, water moves from the solution into the imbibant and *versa vice*. Not only does water pass into dry, mature seeds by imbibition as the first step in germination, but there is evidence that water moves by this process into ovules which are ripening into seeds in the ovulary.

During the latter stages of the development of seeds in the cotton boll, for example, their diffusion-pressure deficits exceed their osmotic pressures. This indicates that an osmotic mechanism does not adequately, account for the movement of water into the maturing seeds during this period. Since no evidence could be found for the existence of a metabolically activated mechanism of water absorption, the most probable explanation of these results is that the diffusion-pressure deficits of maturing cotton seeds are largely imbibitional in origin.

It seems likely that imbibition plays a proportionately large role in the movement of water into many other young, actively growing tissues. Other examples of the imbibitional absorption of water can be seen in certain species native to semidesert regions, such as the resurrection plant (*Selaginella lepidophylla*) or in species such as many kinds of lichens which are indigenous to locally dry habitats such sun-swept cliffs. Such species may become air-dry during periods of drought, but rapidly imbibe water and resume their physiological activities upon the advent of rain.

At least some species of lichens readily imbibe water-vapour as well as liquid water. Within a plant cell the cell sap, the water in the protoplasm, and the water in the cell wall may each be regarded as possessing a diffusion-pressure deficit of its own. The diffusion-pressure deficit of the cell wall is largely of imbibitional origin, and that of the protoplasm partly so. If, for example, the diffusion-pressure deficit of the walls of a mesophyll cell is increased as a result of evaporation during transpirational water loss, water diffuses into the walls from the protoplasm and, in turn, from the cell sap into the protoplasm as long as the diffusion-pressure deficit of the wall exceeds that of the cell sap. In cells in which vacuoles are very small or lacking, or in which the interior of the cell is occupied by hydrophilic colloidal substances, imbibitional phenomena undoubtedly play a greater proportionate role in the movement of water than in cells with prominent vacuoles.

Possible Metabolic Mechanisms of the Movement of Water in Plants

The osmotic and imbibitional mechanisms may not be the only ones which play a role in the movement of water from cell to cell in plants or from the external environment into plant cells. A number of investigators have suggested that metabolic mechanisms may also be involved, it usually being implied that the energy of respiration is utilized in the transport of water molecules across cell membranes. Much of the evidence which has been advances in support of such an hypothesis is highly indirect or otherwise unconvincing. Kelly (1947), however, has shown that a close correlation exist between the rate of respiration and rate of water intake in the oat coleoptile.

It has also been shown that certain inhibitors of respiratory enzymes retard water intake by potato tuber tissue. The demonstration that auxins accelerate water absorption by potato tuber and other tissues when immersed in pure water has also been considered to be evidence for the existence of such a mechanism. Auxins are one type of plant hormone and are known to influence respiration rates. Hence their effect on water absorption might be evidence that respiration plays a role in this process. However, Levitt (1948) failed to find a correlation between the rates of auxin-induced water absorption and respiration in potato tuber tissue, and considers an increased plasticity of the cell walls to be a more likely explanation of this phenomenon.

That correlations do exist between rates of respiration and rates of water movement or absorption in at least some kinds of plant tissues is undeniably true. Whether or not this relationship is a very direct one, however, is not clear. In meristems, for example, passage of water into cells may be a resultant of their increased solute intake, increased fabrication of protoplasm, or expansion in cell-wall area, all processes which appear to be dependent upon respiration as a source of energy. The influence of respiration on water movement may thus be principally or entirely an indirect one. While respiration undoubtedly exerts at least indirect effects on movement of water in plants, there is almost no positive evidence that the energy of their process is utilized directly in the transport of water molecules. Furthermore, according to calculation of Levitt (1947), it appears improbable that a metabolic mechanism of movement of water of any appreciable magnitude could be maintained in most kinds of plant cells at the expense of respiratory energy.

The Pathway of Water through the Root

Water enters the roots principally through the walls of the root hairs and epidermal cells of the root tips. Absorption of water by individual root hairs has been demonstrated experimentally with microphotometers. From the epidermal cells the water passes through successive rows of thin-walled cortical cells, and then through the cells of the endodermis. The structure of the walls of the endodermal cells is peculiar. Two main types of such cells have been recognized. In one type, the inner tangential and radial walls, or sometimes the entire wall, is thickened. These thickened walls are suberized and sometimes partly lignified. In a second type, a thickened strip is present on the inner surface of the radial and transverse walls, this thickening often being suberized. The width and general configuration of these *Casparian strips*, as they are called, vary with the species. Regardless of types most thickened endodermal walls are pitted.

In some species having the thick-walled type of endodermis, there are present, opposite the outer end of each area of xylem tissue isolated thin-walled endodermal cells called passage cells. These are supposed to facilitate the movement of water and dissolved salts through the endodermis. Passage of water and mineral salts through the endodermis is probably also facilitated by the presence of lateral root initials which usually have developed by the time the endoderms is mature. After passing through the endodermis water moves into the xylem ducts, in most species after traversing a few intervening layer of pericyclic cells.

The Relation between Roots and Soil Water

From the standpoint of the absorption of water by plants, a clear distinction should be made between conditions under which capillary movement of water occurs readily in a soil and those under which such movement of water is slow or non-existent. If capillary movement occurs readily, translocation of water may take place toward the young roots whenever they are absorbing water. There are two principal conditions under which capillary translocation of water can occur in soils at appreciable rates (1) in any zone of soil which is not more than a few feet above a water table, and (2) in the upper layers of any soil after a heavy rain or irrigation, but before content of the soil has decreased to its field capacity.

As the water in the films surrounding the soil particles with which the root tips are in contact becomes depleted, more water moves towards those particles by capillarity. The actual rate of such capillary movement of water through the soil may become a factor influencing the rate of absorption. Root systems, however, are not static, but are more or less continually growing through the soil. The rate of root growth of most species decreases, as a general rule, with increasing wetness of the soil above the field capacity, because of the corresponding reduction in soil aeration. Hence in soil in which capillary movement of water occurs – which are necessarily relatively wet – the rate of elongation of roots is generally less than in otherwise similar but somewhat dryer soils. This continued growth of root tips through the soil brings them into contact with other portions of the soil water, so that, even if capillary movement to certain parts of the soil ceases, capillary movement of water to the roots may be re-established by the extension of the root tips themselves into zones of the soil that have not yet have depleted of water which can move by capillarity.

Many plants, much of the time, grow in soils at water contents between the permanent wilting percentage and the field capacity. In this range of soil-water contents capillary movement of water is very slow. Once most of the film water present on the soil particles with which the root tips are in contact has been absorbed it cannot be replaced in any significant quantity by capillary movement from adjacent regions of soil if the soil water content is below the field capacity. Neither does water move in vapour form through soils toward the absorbing regions of roots at appreciable rates.

Under each conditions the absorbing region of every rootlet often becomes surrounded with a narrow cylindrical zone of soil which has been depleted to a water content much below that of the surrounding soil. Since, in soils at a water content below the field capacity, movement of water toward the roots is very slow, the principal method by which the roots come in contact with additional increments of water is by continually growing through the soil. Mature root systems of many species of plants bear millions of root tips. Each of these numerous root tips may be pictured as progressing through the soil and absorbing most of the water present in the smaller interstices between the soil particles with which they come in contact. Relatively large quantities of water can be absorbed in this way, at least by some species of plants. For example, the total length of all the roots on a four-months-old rye plant was found by Dittmer (1937) to be 387 miles.

On the average, therefore, the aggregate daily increase in the length of the roots on this plant was more than 8 miles. In addition nearly 55 miles of new root hairs were formed, on the average, per day. Kramer and Coile (1940) calculated that such a rate of root extension would permit the absorption by such a plant of about 1.6 liters of water daily from a sandy loam, or about 2.9 liters from a heavy clay loam, when both soils were at the field capacity. This picture of the role of root elongation in the absorption of water (and mineral salts) from the soil under certain conditions is probably an accurate one for many and perhaps most species of plants.

The root systems of some species of plants, however, are of a sparse, sparingly branched type which would suggest that the quantities of water which they can absorb in the manner just described would be relatively small. Mycorrhizae may aid in the absorption of water by some species. Elongation of roots would also be less effective in contributing to the absorption of water by terrestrial species in which the roots bear few or no root hairs than in species on which root hairs develop in abundance. The root-tip population of any plant is usually so large that often not all of the root tips are subjected to the same soil-water conditions. Some may be located in lower soil horizons which, under certain conditions, contain more water than the upper layers of the soil. Under other conditions the reverse situation may prevail. After light rainfalls on a comparatively dry soil for example, the root tips closer to the surface may be in contact with soil at a higher water content than those at greater depths. Hence the quantity of water absorbed in a given interval of time may differ greatly from one root tip to another as they advance through the soil. In general however, roots do not grow appreciably in soils which have a water content less than that of the permanent wilting percentage.

Mechanism of the Absorption Water

The development of a diffusion-pressure deficit in the mesophyll cells of leaves causes the water in the xylem vessels or tracheids to pass into a state of tension which results in an increase in the diffusion-pressure deficit of the water in the xylem ducts by the amount of the imposed tension. As soon as the diffusion-pressure deficit of the water in the xylem ducts in the absorbing region of the root exceeds that in contiguous cells, a gradient of diffusion-pressure deficits is established across the root, increasing consistently from cell to cell from its epidermal layer to the xylem conduits. Many authorities believe that the water in the cells in absorbing regions of roots often passes into a state of tension under conditions such as those just described. If this occurs greater diffusion-pressure deficits could develop in the peripheral cell walls of young roots than otherwise would be possible. However, the mechanism as just described will operate even if the water in the root cells never passes into a state tension.

The osmotic pressure of the root epidermal and root-hair cells of most species for which measurement are available is about 3-5 atm., although higher values undoubtedly occur in some species. Hence diffusion-pressure deficits of this magnitude can develop in the peripheral cells of roots even if the water in them is never under tension. Whenever the diffusion-pressure deficit of the water in the peripheral walls of the young root cells exceeds that of the water in soil, water will move from the soil into the root. Since the osmotic pressure of the soil solution in most soils is only a fraction of an atmosphere, the diffusion-pressure deficit of the absorbing cells of a root does not have a to be very great before water will enter them from any soil with a water content equal to or greater than the field capacity. The absorption process which has just been described is often called "passive absorption" because the entry of water into the roots is brought about by conditions which originate in the top of the plant and the root cells seemingly play only a subsidiary role. Although the general picture of this mechanism of absorption which has just been presented is probably correct in its essentials, it is almost certainly oversimplified.

The influence of certain environmental factors upon absorption, particularly temperature and oxygen, suggests that the metabolic activities of living cells in the absorption zone of roots also play at least indirect part in this process. The mechanism of absorption just described undoubtedly accounts for the intake of most of the water which enters the roots of plants, but it is not the only mechanism of absorption which is known to operate in plant. In many species an internal pressure known as root pressure often develops in the xylem. The occurrence of sap exudation resulting from root pressure can be strikingly demonstrated with some species by immersing the root system of a decapitated plant in a potometer. After a time a dilute swap will begin to ooze from the cut stem, and the absorption of water will be indicated by the movement of the meniscus on the capillary arm of the potometer.

If the volume of water exuded is measured it will be found to be not sensibly different from the volume absorbed. In other words, water is being absorbed and is moving in an upward direction through the plant as a result of processes which take place in the root cells. This type of absorption, in which the mechanism involved is localized within the root system, is often called "active absorption." Root pressure, or guttation, and active absorption are usually considered to be different aspects of the same phenomenon. There is good evidence for the existence of a relatively simple osmotic mechanism of active absorption, but it is by no means certain that this is the only mechanism responsible for root pressure, xylem sap exudation, and guttation. The essentials of an osmotic theory of active absorption were first early suggested by Atkins (1916). Priestly (1920, 1922) has also advocated a similar hypothesis. In a number of species it has been shown that, although the osmotic pressure of the sap in the xylem ducts is relatively low, seldom exceeding 2 atm., it is higher than the diffusion-pressure deficit of most soils at the field capacity or a higher water content.

Osmotic movement of water from the soil to the xylem ducts could therefore occur through the "multicellular membrane" of the intervening root cells in spite of the fact that such cells have a higher osmotic pressure than either the soil solution or the xylem sap. The mechanism of such a movement of water can be interpreted in terms of the establishment of a ;consistently increasing gradient of diffusion-pressure deficits from cell to cell across the root from the epidermis to the xylem ducts. Kramer (1932) has shown that in an analogous situation water will move across the cells of the petiole of the tropical papaw (*Carica papaya*). Eaton (1943) found an almost proportional relation between the difference in osmotic pressures of the xylem sap and the external solution and the rate of exudation from decapitated young cotton plants. His results also indicated that exudation would cease when the value of this osmotic differential fell to zero suggesting that, under the conditions of these experiments at least, the osmotic mechanism was the only one in operation.

The sensitivity of the osmotic mechanism is very great ; reversals occur in less than a minute from exudation to cessation of exudation and back to exudation upon transfer of the root systems of young plants from water to a dilute sucrose solution and back water. Although there is little doubt of the existence of an osmotic mechanism of active absorption, the result of certain experiments are difficult to reconcile with the view that this process can be accounted for entirely by such a mechanism. The finding of Grossenbacher (1938, 1939) and Skoog et. al (1938) that an autonomous more or less regular diurnal fluctuation occurs in the rate of exudation of xylem sap from detopped seedlings, with the maximum about noon and the minimum about midnight, suggests a more complex mechanism than the one described above.

A retarding effect of deprivation of oxygen on absorption of water by isolated onion roots, and by root hairs of radish, has been demonstrated by refined experiments with potometers, suggesting that respiration plays at least an indirect part in the active absorption mechanism. The enhancing effect of auxin on xylem sap exudation rates also suggests that respiration plays a part in active absorption, since auxins are known to influence respiration rates. It has therefore been postulated that a second mechanism of active absorption exists which is dependent for its operation upon the metabolic activity of the root cells. The influence of metabolic processes on active absorption of water may, however, be largely or entirely and indirect one.

The absorption of mineral salts by cells is markedly influenced by the metabolic conditions prevailing within them, especially their rate of aerobic respiration. Furthermore, there are good reasons for believing that the passage of salts into the xylem ducts is largely controlled by the metabolic conditions in adjacent living cells. Metabolic conditions, therefore, may influence the steepness of the osmotic gradient between the xylem sap and the soil because of their effect on the absorption and cell-to-cell movement of mineral sales; and their influence on the active absorption of water may be exerted largely or entirely in this or in other indirect ways.

The Relation of Transpiration to the Movement of Water through the Plants

In most discussions of the cohesion of water theory of the rise of water in plants, the relation of transpiration to this process is so greatly emphasized that it has come to be almost indelibly associated with this concept. Transpiration is not however, the fundamental cause of the upward movement of water in plant. Water ascends in the xylem ducts only when an adequate diffusion-pressure deficit has been created in the water of the vacuoles or the walls of cells in organs in the upper parts of the plant. Since evaporation of water from the walls of the mesophyll cells in the most frequent cause of such diffusion-pressure deficits, the process of transpiration has been generally linked in discussions with the mechanism of the ascent of sap. It is only because of its effect in increasing the diffusion-pressure deficit of the water in the mesophyll cells that transpiration sets in motion the entire train of water through the plant.

Any other process which results in an increase in the diffusion-pressure deficit in the cells at the terminus of any plant axis may also induce the translocation of water. Upward movement of water often continues during the night hours after transpiration has virtually caused. This lag results from the residual high diffusion-pressure deficit of the leaf cells at the end of the daylight period. Water will continue to enter these cells until they reattain the maximum turgidity of which they are capable under the conditions prevailing. Similarly movement of water will occur into any rapidly growing stem tip or fruit because the binding up to water in certain in phases of the growth process creates a diffusion-pressure deficit in the cells of such organs, thus inducting the migration of water towards such centers of growth activity. At the present time most plant physiologists are agreed that the cohesion theory is a correct representation of the principal mechanism by which water is transported through plants, be they the tallest of trees or herbs only a few feet in height. It is also generally agreed that root pressure plays a minor role in this process at least under certain conditions and in some species. There is no doubt that living cells are involved in the phenomena generally described under the term root pressure, and there is at least a possibility that living cells of the xylem are in some way essential to the maintenance or operation of the cohesion mechanism.

Rates of Movements of Water through Plants

The rate at which water ascends through the xylem ducts may vary from a movement so slow as to be almost imperceptible to speeds of at least 75 cm per minute and probably higher. In general, daily variations in the rate of ascent of water approximately parallel daily variations in transpiration rate; the maximum velocity of the transpiration stream is often not attained, however, until several hours after the peak of transpiration.

Lateral Movement of Water

A cell to cell lateral movement of water in a radial direction undoubtedly occurs along the vascular rays in the steams of most species of plants. In woody stems there is probably also a lateral movement of water around the stem in a tangential direction. Except in trees in which the "grain" of the wood is twisted, the conducting vessels on one side of the tree generally connect with branches on that side of the tree at their upper extremity, and with roots on the same side of the tree at their lower extremity. If no lateral movement of water occurs in woody stems, it would be expected that removal of the roots from one side of a tree would result in a dearth of water, or perhaps even death of the leaves or branches on that side of the tree. Experiments have been performed on apple, peach, oak, and other woody species in which the roots on one side of the plant were removed in an attempt to ascertain whether or not lateral movement of the water occurs. Although the water content and growth of the plants treated in this manner diminished, there was no difference in the moisture content of the leaves on the two sides of the tree. Neither did the leaves on the side from which the roots had been removed show any greater tendency to wilt on clear, warm days than those on the other side. These results indicate very strongly that lateral movement of water occurs in woody stems, and that the water conductive system of plants acts as a unit system.

Downward Movement of Water

There are a number of experiments on record which indicate that downward movement of water can occur in stems. For example, Dixon (1924) found that, if the tip of the leaf of a potato plant be cut off under an cosin solution, the liquid will descend through the xylem tissue of the leaf, petiole, and stem, and eventually reach the underground organs. In general, however, such an effect is to be expected only when an internal water deficit exists in a plant. The cohesion theory of the movement of water will account equally well for the conduction of water in either an upward or a downward direction through the plant. Whenever conditions are such that the diffusion-pressure deficit of the cells of an organ, which is basally situated relative to another, is greater than that of the cells of a more nearly apical organ, a reversal in the direction of the movement of the water will occur.

Sap Flow in the Maple

Sap flow from cuts or bore holes in maple trees occurs only during the late winter and early spring months when cold nights alternate with days upon which the temperature rises above freezing. The flow is largely or entirely confined to the daylight hours. A single sugar maple tree will generally yield from 25 to 75 liters of sap per season, the sucrose content of which is usually between 2 and 3 per cent.

In other woody species that exhibit a prefoliation sap flow, such as birch and grape, exudation usually will not occur until later in the spring at a period when freezing temperatures have become uncommon. Xylem sap flow in such species is a root pressure phenomenon and continues during both day and night hours. Sap flow in the red maple is not a result of root pressure because sap exuded as freely from borings into the xylem of cut trees with their bases immersed in water as from similar borings into rooted trees during the late winter and early spring whenever the necessary alternation of cold nights and warm days occurred. These investigators suggest that the crystallization of water within the trunk and branches, especially in the outer layers of the xylem, indirectly causes upward movement of water through the tree.

As a result of the conversion of some of the water present into ice, the osmotic pressure of the cells increases, which in turn induces upward movement of water through the xylem. It must also be assumed that at least some of the roots are in soil layers warm enough to permit the absorption of water. When, under the influence of daytime rising air temperatures or direct insolation, the temperature of the trunk and branch tissues reaches or exceeds the freezing point, the ice melts and the tissues become flooded with sap. This sap exudes through any sort of an incision made into the outer xylem of the tree. We know that plants require water; deprived of it, they will wilt and die. In fact, the typical and plant consumes prodigious quantities of water, vastly more than any of the other substances that enter it, and some 100 times more than its content of water at maturity.

Stated differently, the rate of water loss is 100 times greater than the rate of entry of carbon dioxide. Most of this water therefore does not remain within the plant, but passes through it to the atmosphere. The process is called transpiration. The transpiration steam caries to the aerial parts of the plant nutrient ions absorbed by the roots, although the rate of transpiration is not necessarily a limiting factor in the absorption of nutrients. The seemingly wasteful consumption of water through transpiration is nonetheless essential to the growth of plants. In nature and even is agriculture, it is a rare and fortunate plant that enjoys an optimum supply of water throughout its life. Without irrigation, most crops are inhibited, often seriously, from achieving full growth and development.

If we are concerned with the processes controlling the growth of plants in nature or if we are concerned with the worldwide problems of agriculture then we must recognize that there is perhaps no single factor that is more crucial and none more amenable to our intervention (at least in principle) than that of control over transpiration in plants. At one stage, the movement of water in plants involves its diffusion as a gas. We shall therefore want to consider some of the laws of gaseous diffusion, including the behaviour of CO_2 and oxygen as well as water. We shall see that long-distance water movement occurs primarily through a system of capillaries called *xylem*. In contrast, the movement of organic solutes occurs in an independent set of extraordinary channels called *phloem*. We shall examine this process under the term *translocation*.

The Path of Transpiration

Let us first survey the phenomena of transpiration, looking especially at the anatomy of water movements. We shall here take as facts certain notions which we may wish to examine more critically in the sections that follow. Water is absorbed from soil through young roots, especially by root hairs. The water appears to move through the cortex of roots and through their cell walls. Gonical cells are closely interconnected by strands of protoplasm called *plasmodesmata*. The collection of interconnecting protoplasm is often called the *symplast*.

The peculiar anatomy of the endodermis, composed of cells with radially thickened and suberized cell walls, has led many physiologists to infer that it is a barrier to water movement, that water at this point must pass through the cells rather than the cell walls. Although this notion is of only marginal relevance to the problems of water movement, it becomes of major importance for the leading hypothesis on ion transport. Whatever are the more subtle aspects of radial movements of water across the root, once water enters the stele its distribution through the plant appears to be a matter of plumbing, rather elegant plumbing but nonetheless plumbing.

Pteriodophytes, gymnosperms, and angiosperms evolved conductive cells called tracheids. These cells are elongated, ranging typically from a few tenths of a millimeter to a few millimeters, and exceptionally to 10 or 20 mm in length, with strongly lignified walls. Since water must move across many more cross walls in tracheids than in vessels, the efficiency of water conduction is substantially less. The inefficiency is somewhat compensated by frequent lateral connections, known as bordered pits. These pits appear to be cleverly designed check valves, which may be able to seal off a tracheid which has suffered an embolism.

Angiosperm wood, in addition to tracheids, also contains the hydrodynamically more efficient *xylem vessels*. These structures may be up to several meters long and a few tenths of a millimeter in a diameter and can conduct impressive volumes of water. The vessels are the mortal remains of vertical chains of cells whose lateral walls are sealed with lignin and other amorphous polysaccharides and strengthened with helical thickenings of cellulose. The endoplates have dissolved away to form long capillary channels. The protoplasm has disappeared. Water in vessels is typically under negative pressure. Since the introduction of bubbles into these capillaries will destroy the conductive system, the capillary column will often be broken in winter when the water freezes.

In order to succeed in so-called temperate climates, dicots learned simply to abandon air-filled vessels and cut off new vessels from the cambium each spring before new leaves expand. All along the conductive system a modest amount of lateral transport supplies the living cells of the stem with inorganic nutrients and replenishes the relatively small amounts of water lost by evaporation. Most of the transpiration stream continues upward to the leaves. There, small amounts of water and the remainder of the nutrients enter the leaf parenchyma. The bulk of the water passes along the parenchyma cell walls and evaporates into the intercellular spaces. Finally the water diffuses out into the atmosphere through the *stomata.* The water system of a tall tree or liana (woody vine) is an astonishing device, especially when viewed in developmental sequence. The water columns are first filled by capillarity and osmosis as the living protoxylem cells expand.

The water remains as the end walls dissolve and the protoplasm disappears. The special meristems and leaves are carried tens of meters above their source of water as the plant grows. At their upper ends the capillary columns "hang" on the evaporative surfaces of the leaves. Consequently the columns are under such enormous negative pressures (tensions) that if the tiniest bubble forms, the columns snap and cease to conduct. Nonetheless, the water system remains in a kind of secular equilibrium; old vessels lose their function only to be replaced by new ones periodically formed from the cambium. We have said that the transpiration stream brings inorganic nutrients to the aerial parts of the plant, that the volume of water transpired is larger than required for this purpose, and also the most of the water evaporates through the stomata.

It is obvious that, if the rate of water loss is not subject to some kind of control, the plant's tissues will suffer from desiccation, as would those of any organism. Even though the leaves and stems are enveloped in a waxy cuticle which is almost impermeable to water, the stomata are so ingeniously designed that, when they are open, water can evaporate through them almost rapidly as if the entire leaf surface were open to the atmosphere! The seeming death wish on the part of plants becomes understandable when we recognize that the stomata are programmed to admit carbon dioxide for photosynthesis. Paradoxically in order to carry on maximal rates of photosynthesis, plants must accept a potentially dangerous loss of water. Land plants are thus forever poised between the Scylla of carbon starvation and the Charybdis of desiccation !

Considering the seeming limitless capacity of plants for chemical synthesis, we can only conclude that the failure to plants to develop an epidermis that would admit carbon dioxide but retain water vapour is some cosmic oversight. In the sections below we shall inquire more rigorously into the movement of water in plants. We shall then consider as case studies some of the limiting factors governing water movements in the several stages of transpiration.

Hydrostatics and Equilibrium Relations of Water in Plants

Exactly as in the case of chemical transformations net movements of water can occur only in the direction of a decrease in the free energy of the system. We must therefore inquire into the factors within the soil-plant-atmosphere system which may affect the free energy of water. Although investigators concerned with soil physics have reported states of water in terms of its free energy, plant physiologists have usually used term as "diffusion pressure deficit," which are difficult to relate rigorouly to thermodynamic quantities. Recently there has been general agreement to use the them water potential, ψ defined as the chemical potential per unit molal volume of water.

$$\psi \equiv \frac{\Delta \mu w}{V} \qquad \text{...(1)}$$

where V is the partial molal volume of water (18 cc $mole^{-1}$) and Δμw is the difference between the chemical potential of water in the solution and the chemical potential of pure water at 0° and 1 atm pressure. Chemical potential has its usual thermodynamic meaning of the change in free energy per mole at constant temperature and pressure (Equation 2)

$$\mu w \equiv \text{chemical potential of water} \equiv \left(\frac{\partial G}{\partial n}\right)_{T,P} \qquad \text{...(2)}$$

The term water potential at once keeps physical chemists content and has intuitive appeal to biologists; as with other potentials, water will tend to flow from higher to lower water potentials. Similarly, the water of solutions with identical ψ values will be in equilibrium ; that is, the net flux will be zero. Water potentials can be thought of as compared of three terms, attributable to osmotic pressure, hydrostatic pressure, and the so called matric potential or matric effect (Equation 3).

$$\psi = \psi o + \psi_P + \psi_M \qquad \text{...(3)}$$

ψ and the terms that compose it have the units of pressure. Under standard conditions (pure water at 0° and 1 atm), ψ = 1 atm. We shall give here a brief account of how each term is generated.

ψo and Osmotic Pressure

If a solution contains solutes, the water potential will be less than that of pure water. This happens quite simply because the presence of solutes decreases the concentration of water itself. More precisely, it is the activity of water which is decreased. (Activity is another of the useful fictions invented by physical chemists : it is the concentration of a solute or solvent in an ideal solution.) Quantitatively, the decrease in water potential due to solutes, ψo, is related to the activity of water by Equation 4.

$$\psi o = -\frac{RT}{\overline{\overline{V}}} \ln \frac{a^\circ}{a} \quad ...(4)$$

where R is the universal gas constant, T the absolute temperature, V the partial molal volume of water, and a° and a the activities of pure water and the water in the solution.

ψ is also the negative of the osmotic pressure π.

$$\psi^\circ = -\pi \quad ...(5)$$

If the solution is "ideal"

$$\psi o = -\frac{RT}{\overline{\overline{V}}} \ln \frac{1}{N\omega} \quad ...(6)$$

where Nω is the mole fraction of water.

For solution which are both dilute and ideal,

$$\ln \frac{1}{N\omega} \approx \Sigma n_S$$

Where n_s is the number of moles of solute. This leads to the van't Hoff relation.

$$\pi = \frac{\Sigma n_S RT}{\overline{\overline{V}}} \quad ...(7)$$

Note that since the osmotic pressure π must : always be positive, the ψo term is always negative.

ψ_P and Hydrostatic Pressure

The pressure term, ψ_P, is numerically equal to the hydrostatic pressure.

$$\psi_P = P \quad ...(8)$$

Hydrostatic pressure may be either positive or negative. The inward pressure on a cell exerted by the cell wall is called the turgor pressure and is positive. Negative pressure or tension can exist in xylem elements.

ψ_M of Matric Potential

The peculiar properties of water are largely accountable for tendency of water molecules to form hydrogen bonds with other water molecules, with solutes, and with hydrophilic colloids. Water has a large electric dipole moment, so that the electron-rich (oxygen) region of water molecules tends to align toward electron-poor (cations), while the hydrogen ends are oriented around anions. Ions typically attract a shell of water up to 2Å deep, depending on the electric field strength. The shell is called water of hydration. The hydrogen bonding of water with oxygen-, nitrogen-or sulphur-containing radicals of large molecules, such as proteins and polysaccharides, may be extremely extensive.

The association of water with large molecules results in the phenomenon of *imbibition*, whereby seeds absorb large amounts of water even from relatively dry soils. Finally, water may be bound in tissue capillaries by van der walls' forces. All of these forces tend to lower the potential of water and are lumped as ψ_M The determination of ψ_M is largely empirical and is measured as the water potential after the hydrostatic pressure and solute effects have been eliminated or substracted out.

Calculations Involving Water Potentials

In order to predict the directions or rates of water movements in the soil-plant-atmosphere system, we should gain some experience in calculating water potentials of various plant compartments.

Example 1. What is the water potential of a 1 molal solution of sucrose at 25° and 1 atm pressure ?

Let us assume that a 1 molal solution is "ideal." Starting with Equation 3.

$$\psi = \psi o + \psi_P + \psi_M$$

we may take as given that ψ_P and $\psi_M = 0$.

From Equation 6.

$$\psi o = -\frac{RT}{\overline{V}} \ln \frac{1}{N\omega}$$

$$R = \frac{18.06 \text{ cc atm}}{\text{deg mole}} \quad T = 298° \text{ K}$$

$$\overline{V} = 18.07 \text{ cc}$$

$$N\omega = \frac{55.5}{56.5}$$

(that is, the mole fraction of water in the solution is equal to the number of moles of water divided by the total number of moles present).

$$\phi_o = -\frac{82{,}06.298}{18.07} \ln \frac{55.5}{56.5} \text{ atm}$$

$$\psi = \psi_o = -24.5 \text{ atm}$$

Example 2. What hydrostatic pressure is required to cause the water potential of water in 1 molal solution of sucrose to equal that of pure water at 25° and 1 atm ?

We want ψ_P. We determined in Equation (1) that $\psi o = -24.5$ atm. Since $\psi_M = 0$, we have

$$0 = -24.5 \text{ atm} + \psi_P + 0$$

$$\psi_P = 24.5 \text{ atm}$$

Hence, the pressure required would 24.5 atm.

Note: The hydrostatic pressure required to bring a solution into equilibrium with pure water is defined as the osmotic pressure of that solution.

Electricity of Cell Walls

Since the ψ of root cells is generally less than that of soil water, water tends to enter the roots cells. The root cells swell and hydrostatic pressure develops as the cellulose walls are stretched. This is called turgor, and the hydrostatic pressure is referred to as turgor pressure. The cells walls are neither perfectly nor infinitely elastic so that the change in cell volume is a complex function of the external water potential.

Example 3. The volume of a cell is measured, and then the cell is placed in water at 25°, in which it is observed to swell. It is then placed in successively increasing concentrations of mannitol; each time it shrinks some. When it is put into 0.2 M mannitol it is found to be the same size as it was initially. It continues to shrink in stronger solutions, until in 0.5 M mannitol very slight plasmolysis is seen, while 0.6 M mannitol the volume of the cell is 5 per cent less than it was initially. Calculate, in atmosphere:

(a) The cell's ψ_P in its initial state
(b) The cell's ψ in its initial state
(c) The cell's ψ_0 in its initial state
(d) The cell's ψ_0 at zero turgor

As the cell is placed in progressively more concentrated mannitol, water will begin to leave the cell, causing a decrease in volume until the ψ_0 of the external solution reaches the ψ_0 of the cell. At this point the elastic cell walls begin to relax; the hydrostatic pressure ψ_P falls. At zero turgor where $\psi_P = 0$ and (ignoring ψ_M) = ψ_0 (outside) = ψ_0 (inside), the cell contents will just begin to pull away from the cell walls. This is called *incipient plasmolysis*.

(a) In the example, incipient plasmolysis occurs in 0.5 M mannitol; hence, from Equation (7)

$$\psi_0 = -0.5 \times 0.08206 \times 298 = -12.2 \text{ atm}$$

(b) Since the cell volume is 0.5 M mannitol is 95 per cent of the initial volume, it follows that the initial ψ_0 must have been

$$\psi_0 = -0.95 \times 0.5 \times 0.38206 \times 298 = -11.6 \text{ atm.}$$

(c) The effect of 0.2 M mannitol is equivalent to the original hydrostatic pressure; hence,

$\psi_P = 0.2 \times 0.08206 \times 298 = 4.9$ atm.

(d) The initial $\psi = \psi_0 + \psi_P = -11.6 + 4.9 = -6.7$ atm.

Air

The calculation of water potentials of water vapour in air in similar to the situation in solutions. We substitute the notion of partial pressure of the gas for the concentrations. (More precisely, we should substitute *fugacity*, the idealized partial pressure for activity in Equation 4).

$$\psi_0 = -\frac{RT}{\overline{V}} \ln \frac{p^o}{p} \qquad ...(9)$$

where p° is the partial pressure in equilibrium with pure water and p is that for the solution in question.

Example 4. Calculate the water potential of air at 90 percent relative humidity at 25°.

$$\psi = \psi_0 + \psi_P + \psi_M$$

$$\psi_P = \psi_M = 0$$

$$\psi_0 = -\frac{RT}{\overline{V}} \ln \frac{p^o}{p}$$

$$\psi_0 = -\frac{82.06 \times 298}{18.07} \ln \frac{100}{93} \text{ atm}$$

Hence

$$\psi = \psi_0 = -141 \text{ atm}$$

The astonishingly low water potential of even moist air means their equilibrium during the day between water in the leaf and in the air is virtually impossibly.

Hydrodynamics: Laws Governing the Rate of Fluid Movement

We considered the factors governing water potentials in and around plants. Differences in water potential will determine the direction taken by water movements, but exactly as in the case of chemical transformations, the rate of water movements is determined by additional factors. The movement of liquids in capillaries is well understood. From Poiseuille's law (Equation 10), we can predict rates of volume flow J.

$$J_V = \frac{\pi r^4 \Delta P}{8 \eta l} \qquad ...(10)$$

given the radius of the capillary, r; the viscosity η; and ΔP/l, the mean pressure gradient along the capillary. From the r^4 factor we may infer that in comparison to a tracheid of 10μ radius, a vessel with a radius of 100μ would have 10^4 greater specific conductivity!

We can similarly predict the flow of gases through stomata from Fick's law (Equation 11)

$$J = -DA\frac{\partial s}{\partial x} \text{ or } \frac{J}{A} = -D\frac{\partial s}{\partial x} \quad ...(11)$$

where D is a diffusion coefficient, $\partial s/\partial x$ is the concentration gradient of a solute s with respect to distance x, and A is the cross-sectional area.

Serious difficulties arise when we come to a membrane. The property of membranes to pass fluids is called *permeability*. Nearly all substances move in or out of cells less rapidly than would be expected on the basis of free diffusion. Some substances (for example, proteins) appear to move across cell boundaries at essentially zero rates. We conclude that cells and tissues posses restricted permeability. We say that membranes are differentially permeable or, ambiguously, that they are "semipermeable." Let us imagine an idealized membrane of thickness Δx separating an outer solution I from an inner solution II. The solutions contain water w and solute s. The chemical potentials μi of the water and solute are shown as μw^I μw^{II}, μ_s^I, μ_s^{II}, etc.

Intuitively (that is, viewing the system as simple plumbing), we can imagine characterizing the membranes in part by the rate of flow of fluid as a function of the pressure drop across the membrane. Let J, ≡ total flow across the membrane. If the osmotic pressures of I and II are equal, we could expect that flow would be simply proportional to the pressure drop ΔP:

$$J_v = L_P \Delta P \quad ...(12)$$

where we define L_P as the hydraulic or filtration coefficient for unit pressure difference. We can also imagine characterizing the membrane by the rate of diffusion of solute across the membrane. With zero electrical potential difference, we expect according to Fick's law (Eq. 11)

$$J_S = -\omega\frac{ds}{dx} \quad ...(13)$$

where ω is the diffusion coefficient for the solute and ds/dx is the concentration gradient across the membrane. Assuming a homogeneous membrane, ds/dx should be (intuitively) proportional to the concentration difference, SI—SII. Hence Equations (13) could be rewritten

$$J_S = -k(\pi I - \pi II) \quad ...(13)$$

where k is an arbitrary constant.

Both of these kinds of measurements are of limited application since flows arise in the intact plant from differences of both osmotic pressure and hydrostatic pressure acting simultaneously. The problem was attacked intuitively by ursprung and Blum, who hypothesized that the rates of flow across membranes are proportional to differences in water potential (Equation 14)

$$J_v = k(\psi I - \psi II) \quad \text{....(14)}$$

or ignoring matric potentials

$$J_v = k[(PI - \pi I) - (PII - \pi II)]$$

This same relation is known in animal physiology as Starling's hypothesis. Unqualified acceptance of Equation 14 resulted in certain distressing contradictions, which will be discussed below. Since Equation 14 was not derived rigorously, biophysicists have searched for a theoretical approach to permeability based on thermodynamics. Katchalsky (1961) assayed the problem as follows:

"Modern research on biological permeability is mainly concerned with the molecular organization of the membranes. The ultimate goal of the investigation is the development of structural models which can be used for prediction of the physicochemical properties of the membranes."

"During the last few years, another approach is emerging whose main concern is not the construction of adequate models, but the establishment of equation capable of organizing the experimental results in a consistent formal framework. This approach is based on the thermodynamics of irreversible processes. The thermodynamic approach admittedly cannot yield quantitative numerical predictions as derived from the treatment of a tangible model. However, it has the advantage of being general, free from contradiction and able to predict new correlations. There is, moreover, a possibility of bridging the two approaches by expressing the thermodynamic few parameters in terms of physical coefficients, such as frictional coefficients, amenable to kinetic interpretation".

The following is an account of the basic transport equation greatly abbreviated from Karchalsky. According to the thermodynamics of irreversible processes, the rate of flow of water is

$$J_w = L_{ww}X_w + L_{ws}X_s \quad \text{...(15)}$$

The symbols deserve some explanation: X_w and X_s are the gradients of chemical potential (cf. Equation 2 across the membrane).

$$X_W \equiv \frac{-d\mu w}{dx}$$

$$X_S \equiv \frac{-d\mu_S}{dx}$$

L_{ww} and L_{ws} are called Onsager coefficients. It seems intuitively reasonable in Equation 15 that the flow of water should be proportional to the gradient of chemical potential of water. What we are unprepared for in the dogma of irreversible thermodynamics is that the flow of water may also be affected by the gradient of solute potential. I say may because L_{ws} factor may be positive but it may also be zero.

By the same token, the flow of solute across the membrane, Js, is also the resultant of two terms.

$$J_s + L_{ss}X_s + L_{sw}X_w \quad \text{.... (16)}$$

From Equation 16 we see that the movement of solutes is a function of the solute potential gradient X_s but may also be affected by the gradient of water potential X_w.

According to the disciples of irreversible thermodynamics, the necessary and sufficient characterization of any membrane would consist in a specification of the coefficients L_{ww}, L_{ss}, and L_{ws} ($=L_{sw}$) for each solute. In dealing with water movements we should like to use the same terms P and π, which proved useful for water equilibria. We can write a pair of equations analogous to Equations 15 and 16, but with P and π substituting for X_s and X_w as the driving forces. J_w and J_s must also be replaced with J, and J_D.

$$J_v \equiv \text{total flow across the membrane} = L_p\Delta P + L_{pD}\Delta\pi \quad \text{...(17)}$$

$$J_D \equiv \text{relative flow of solute vs. water} = E_{pD}\Delta P + L_D\Delta\pi \quad \text{....(18)}$$

where ΔP is the pressure difference and $\Delta\pi$ is the difference in osmotic potential across the membrane, L_p, L_{pD}, and L_D are a corresponding set of onsager coefficients.

We can appreciate the physical significance of the new onsager coefficients by letting the two forces becomes zero separately. When $\Delta\pi$ is zero, J_v is the flow resulting from only a pressure difference, and L_p is seen to be the same coefficient of filtration that we encountered earlier.

$$J_v = L_p \Delta P \quad \text{...(12)}$$

when $\Delta P = 0$, J_D is the exchange flow, so that L_D is the coefficient of diffusion across the membrane.

$$J_D = L_D \Delta\pi \quad \text{...(19)}$$

The significance of the cross coefficient L_{pD} can be seen in Equation 17 when $\Delta P = 0$. Then $J_v = L_{pD}\Delta\pi$ which means that L_{pD} is the coefficient for volume flow with unit osmotic gradient.

Another way of seeing the significance of L_{pD} is simply to adjust the pressure so that there is no volume flow; that is, $J_v = 0$. Then

$$\frac{\Delta P}{\pi} = -\frac{L_{pD}}{L_p} = \sigma \qquad ...(20)$$

But for an ideal membrane $P = \pi$ under equilibrium conditions; therefore, σ is the ratio of the *apparent* osmotic pressure to the theoretical osmotic pressure. The ideality of a membrane is thus expressed by σ, the reflection coefficient.

$$\sigma \equiv -\frac{L_{pD}}{L_p}$$

The traditional (that is, ideal) differentially permeable membrane will have unit volume flow for unit osmotic gradient, so that $\sigma = 1$. But for a completely nonselective membrane under these conditions, osmotic flow is zero, and hence $L_{pD} = 0$. For real membranes, therefore, σ varies between 0 and 1. Equation 17 may now be rewritten.

$$J_v = L_p(\Delta P - \sigma\, \Delta\pi) \qquad ...(21)$$

This leaves L_D as the only coefficient which cannot be conveniently measured. It turns out that the solute permeability coefficient ω is related to the Onsager coefficient by Equation 22.

$$\omega = \frac{L_p L_D - L_{pD}{}^2}{L_p} \bar{C}_s \qquad ...(22)$$

where C_s is the mean concentration of solute in the membrane. Thus a membrane can (in principle) be completely characterized by two conventional coefficients and a new one arising from the Onsager equations.

L_p = hydraulic or filtration coefficient
ω = solute permeability coefficient
σ = reflection coefficient

Evaluation of Membrane Properties and the Idea Semipermeable Membrane

The hydraulic coefficient can be evaluated fairly directly for the relatively spacious membranes of giant algae, such as the *Characea*, but with much greater difficulty and uncertainty in cellular organisms. Measurements of the solute permeability coefficient ω have also been made, again most directly with giant algal cells. The permeability of membranes toward nonelectrolytes follows a very curious rule, first identified by Overton, an English physiologist working in Switzerland in the last century.

According to Overton's rule, natural membranes behave as if they contained a layer of lipid. Since a substance could traverse the membrane only by traversing the lipid layer, the permeability is related to the lipid solubility of the substance, more specifically the distribution coefficient between lipid and water. Lipid solubility turns out not to be the only factor in permeability. Molecular weight also influences permeability to about the 1.5 power. Collander has determined the permeability of the cell membrane of *Nitella*, one of the *Characea*, toward a large number of substances. One of the striking exceptions to the Overtone rule is the permeability of all membranes toward water.

The observed values are as much as 1,000 times greater than would be predicted from the Overton rule. A determination of the reflection coefficient σ will indicate the ideality of a membrane. In an idea semipermeable membrane,

$$\sigma = 1$$

$$L_{pD} = -L_p$$

In such a case the volume flow J_v in Equation 17 becomes

$$J_s = L_p(\Delta P - \Delta\pi) \qquad \text{...(23)}$$

or from Equations 1, 5, and 8

$$J_s = L_p\,\Delta\psi$$

This states that volume flow is simply proportional to the difference between the pressure and osmotic gradients across the membrane. This of course is our old friend, Equation 4 from Ursprung, Blum, and Starling.

It was only reasonable that a gradient of osmotic pressure should cause difference in the activities of water on the two sides of a membrane and hence a net diffusion of water from higher to lower activity. But it was not obvious how osmotic pressure would generate a flow equivalent to that produced by hydrostatic pressure. One can define an osmotic permeability coefficient p_o for water as

$$P_0 = \frac{L_P RT}{V_0}$$

with the units of cm sec^{-1}.

If osmotic flow occurs only by diffusion, the coefficient for the movement of water due to diffusion can be related to hydraulic flow by the equation

$$\frac{L_P RT}{V} = P_d = P_o \qquad \text{...(24)}$$

where P_d is the diffusion permeability coefficient of the membrane for water. Attempts were made, therefore, to test Equation 23 through the prediction of Equation 24 P_0 was evaluated by measuring the bulk flow of water induced by differences in osmotic pressure ; P_d was evaluated by measuring the movement of labeled water, DHO or THO, into or out of cells that were at hydrostatic equilibrium. The results showed that P_0 exceeded P_d by a factor of 10. Solomon, Ussing, and others argued that these results could be rationalized if the transport of water (generated by ΔP or $\Delta\pi$) occurred not by diffusion alone but also by bulk flow through microcapillary pores in the membrane. It is possible that pores may be detected by electron microscopists armed with greater resolution, but there is a simpler solution: it may be that membranes tested are not ideal that $\sigma < 1$. Doubts of the existence of these pores are based on two lines of reasoning: (1) Equation 21 stands on thermodynamic not empirical or initiative, foundations, so that no special mechanism is needed in "explain at"; (2) the existence of unstirred layers of solution on the order of 10μ thick is sufficient to make calculation of P_d extremely doubtful. Such unstirred layers are far less likely to exist under conditions of bulk flow.

In fact, the one known case where P_0 equals P_d is that of *Valonia*, a giant unicellular alga. The dimensions of the alga provide two principal advantages; (1) the effects of the unstirred layer will be small compared with the dimensions of the cell, and (2) the cells can be perfused internally, which facilitates the measurement of the permeability of rapidly permeating substances. Thus, where Equations 23 and 24 can be evaluated with confidence, the results are consistent with a simple, non perforate membrane model. The same problem of unstirred layers prevents the accurate estimation if σ-values. Thus until improved techniques are at hand, e.g., until we can object isolated membranes to direct physical chemical analysis, the full characterization of natural membranes remains incomplete.

Membrane Structure

The molecular structure of membranes is also disputed. There are two extreme models—the lipid and which or lipid bilayer of Davson, Danielli, and Robertson and the lipoprotein subunit hypothesis of Benson. The lipid bilayer model finds support in electron microscopy where membranes can sometimes be visualized as a triple layer of two 20-Å electron-dense layers, thought to be protein, enclosing a 35-Å electron-transparent layer, thought to be lipid. But the optical rotatory dispersion spectra of a number of natural membranes show substantial helical structures among the proteins which is inconsistent with the spread form or monolayer predicted from the bilayer model. Benson finds support for his alternate view of lipoprotein subunits in the pock-marked appearance of freeze-etched chloroplast membranes.

Sjostrand, Green, and Perdue, in an extension of the lipoprotein subunit model, visualize a two dimensional sheet of lipoproteins in which different proteins are imbedded in the same sets of lipids, forming a functional mosaic. It is safe to conclude that the several-models of membrane structure remain unproven.

Kinds of Evidence Required for the Characterization of Fluid Movements in Plan

Since we cannot at this time expect to achieve the kind of characterization of membrane properties required by thermodynamics, the best we can do is a semiquantitative treatment as follows:

1. Measurement of parameters, e.g., vessel radii affecting fluid movements through nonmembranous paths, and factors affecting these parameters.
2. Information on the variables affecting water potential in a given plant compartment.
3. Data on the anatomical pathways of fluid movement, including the ultrastructure of the membranes, plysmodesmata, and cell walls involved.

Case Study : Water Movement in Soil and Roots

The water potential of soils depends in a characteristic way on the water contents of the soil. The relation is approximately hyperbolic. We may note, for example, that the water content of sand must be decreased to one-tenth of *field capacity* before the potential falls appreciably, whereas the water potential in the clay soil begins to fall at about one-half of field capacity. Clearly, substantial amounts of water are bound by clay. The "permanent wilting percentages" of the soils are also noted. At first blush we might imagine that at these values the water potentials of water in the soil and in the roots are equal.

Roots of mesophytic plants in fact have water potential down to about — 7 atm. which is much less than the ψ of soils at the "permanent wilting percentage." Clearly some factor or factors other than water potential determine the point of irreversible wilting of plants. We presume that the rate of water movement into roots follows Equation 23 (or Equation 21 if the membranes are not truly differentially permeable). From this equation we can deduce some of the physiological variables that should govern the movement of water into roots and test if roots in fact behave in this way. The movement of water into roots should vary proportionally with the area of the absorbing surface, A and inversely with the path length *l*. Furthermore, if the membranes were ideal, water movement should be proportional to the difference in water potential:

$$\Delta\psi = \psi_{roots} - \psi_{soil} = P_{roots} - \Delta\pi$$

The system at first blush seems enormously heterogeneous : different parts of the root may have different membrane properties, different π, and different P. Moreover the external surface is not the only, and perhaps not even the, limiting barrier to water movement. From any one point on the root surface, water on its journey to the xylem may have to pass a dozen or more membranes with different parameters. However, certain generalizations are possible : root hairs comprise a major fraction of the absorption surface. These structures seem to be most abundant under moist conditions, when they are needed least.

But in plants that bear root hairs, they account even at the minimum for over 60 percent of the total root surface. It seems agreed that maximum water absorption takes place on the parts of the root where root hairs are most abundant. This region also concides with that of xylem differentiation and lack of suberization. The prodigious extent of roots is almost legendary. Annual crop plants in good soil typically penetrate to 2 m depth; trees and some xerophytes may penetrate to 10 m more.

Table 2.3. Rates of water movement into roots of various plants.

Material	*Rate mm ³ cm ² hr -1*
Corn, young roots in water	20.0
Sour orange, suberized roots in water	5.0
Onion young roots in water	50.4
Radish, root hairs	33.3
Corn, root hairs	28.3
Coffee tree, entire root system in soil	0.25
Short leaf pine (*Pinus echinata*), attached, suberized roots in soil	2.63
Short leaf pine, attached, roots in water	3.37
Short leaf pine, excised, suberized roots, 0.4 atm tension	9.0
Dogwood (*Cornus florida*), same as above	15.6
Yellow poplar (*Liviodendron tulipfera*), same as above	101.4

The general applicability of Equation 17 can be tested with the system. Brouwer (1953) tested *Victa faba* roots with one such system in which the water deficits of the medium and the tissue could be varied independently. He found that the conductivity was constant with varying water potential of the medium but was not constant with changing internal water potential. This is remarkably general finding Increased water deficits, normally resulting from increased transpiration, depress somewhat the conductivity of tissue near the tip but greatly increase the conductivity of order tissue further from the tip. The result is that with increased transpiration, there is a shift in the region of maximal water absorption toward order tissues.

Water conductivity is also strongly depressed by anaerobiosis and cyanide poisoning. Hence conductivity contains metabolically sensitive components. If we reflect on the nature of water conduction, we may inquire what might constitute the metabolically sensitive elements in this expression. There are several possibilities : metabolism may control the path length (thickness of the membrane) or the membrane characteristics. Of the terms that compose the water potentials of root tissues, the hydrostatic pressure P is typically more important than the osmotic pressure $\Delta\pi$. Because of evaporation from the leaf surfaces and the development of large water tensions in the *xylem*, P is normally negative.

In contrast, $\Delta\pi$ (and hence—ψo) is rarely larger than 6 atm. The importance of P can be seen by comparing the rates of water absorption in intact with the 100-fold lower rates of de-topped plants. The conductivity of water in roots is very sensitive to temperature. In fact the water movement through killed *Helianthus* roots at different temperatures almost exactly parallels the viscosity of water. Living roots show a steeper response to temperature, especially in the case of warm-weather plants. That there should be species differences in the response to temperature strongly supports the notion of a metabolically controlled component in conductivity.

The correlation with viscosity reminds us of the microcapillary model for bulk flow through membranes. Paradoxically, we may think of the roots as organs of water absorption but equally well as barriers to absorption. Indeed, below some critical ratio of root to leaf area, transpiration decreases linearly with root area but if the roots are cut off and the stems placed in water, transpiration is more rapid than in the intact plant. Thus, in sum, it seems that roots are more than merely pieces of plumbing, but along with the stomata, roots are carefully controlled devices designed to respond to adverse conditions of water supply.

Case Study : Water Movement in Stems

The water potentials of stem tissues are lower than roots, but of the same order of magnitude. The striking characteristic of water in stems, particularly of tall vines arm trees, is the existence of *negative pressure* (tension) in the xylem. Since xylem sap is fairly dilute (that is, has a small ψo) and since it must be in near equilibrium with surrounding tissues, the hydrostatic pressure must become negative as the water potentials of these tissues fall. The conductive tissue of xylem consists of tracheids and vessels. These structures resemble capillaries, and we may consider the physics of water movement through them accordingly. The factors governing the rate of movement of water through capillaries are expressed in Poiseuille's law (the same as Equation 10).

$$J_V = \frac{\pi r^4 \Delta P}{8\eta l} \quad ...(25)$$

where η is the viscosity, $\Delta P/l$ is the mean pressure gradient along the capillary, r is the radius of the capillary, and l is the length.

Table 2.4. Efficiency of water conductivity in various plants. Comparison of the observed rates of water movement (V_o) comparcd with rates predicted (v_i) from Poiseuille's law.

Plant	*Conductivity Efficiency* $v_o/v_i \times 100$
Vitis vinifera	100
Aristolochia sipho	100
Atragene alpina	100
Root wood of oak	84
Root wood of beech	37.5
Helianthus annuus	32
Rhododendron ferrugoineum	20
Rhododendron hirsutum	13

As we noted above, a vessel with a radius of 100μ would have a 10^4 greater specific conductivity than a tracheid of 10μ radius would have. But does the water in xylem obey Poiseuille's law? We restate Equation 25 as the sum of all conducting elements in a stem:

$$J_V = \frac{\pi}{8\eta}\frac{\Delta P}{l}\sum_{i=1}^{i=n} r_i^4 \quad ...(26)$$

The water conduction of the wood of lianas (first three species) is precisely that expected. The conductivities of oak wood are nearly equal to theoretical, whereas those of the other species range down to a small fraction of the expected values. Generally close agreement between predicted and observed rates of flow through the xylem of tomato was also observed by Dimond (1966).

Deviations from Poiseuille's law should be anticipated tracheids and vessels have end walls. Short vessels and especially tracheids must show conductivities less than that calculated for an open-ended tube. The lengths of tracheids are fairly easy to determine in longitudinal sections. Vessels present more of a problem. Huber's method (1956) is to macerate a given length of woody tissue and count the percentage of end walls. This procedure is directly analogous to end-group analysis in carbohydrates and proteins. Another is to measure the rise of mercury into the capillaries of the wood. Scholander *et al.* (1957) employed still another method with lianas, when segments of the vine were cut at progressively lower levels, water was sometimes released from the lower end. The release of water occurs when a vessel is cut for the first time. From the volume released and the diameter of the emptied vessels, the lengths of the vessels can be calculated. It has even been proposed that in some plants, single vessels may extend from the roots to the leaves.

Table 2.5. Typical lengths of tracheids.

Plant Type	*Length, mm*
Medullosac	24
Araucaria	5(1.5 – 9)
Conifers	2-3
Populus	1
Fraxinus	0.12–0.29

Huber (1956) calculated an empirical quantity called the "relative conduction surface". It is the ratio of the area of the conducting elements to the fresh weight of the plant. A very large fraction of the values fall between 0.2 ant 0.8. Since the conductivity is related not to the area but to the square of the area of the xylem elements, the degree of constancy of the relation is somewhat surprising. We should also remember that in general only the xylem elements formed by recent growth are fully filled with water; the remainder are stopped by air "embolism". Therefore the measurement of total capillary cross sections overestimates the actual conduction area available to the plant.

An ingenious method for the measurement of the velocity of the transpiration streams employs a localized heating of the plant sap together with thermocouples for measuring the placement by the transpiration stream of increased temperature along the stem.

Zimmermann (1964a) employed a variant of this scheme to prove that there was no interruption in water conduction in wood at temperatures in the vicinity of 0°, provided that no actual freezing occurred.

Table 2.6. Typical lengths of vessels

Type of wood	*Length, mm*
Diffuse-porous wood	80-150
Fagus	200
Ring-porous wood	500-1800

Movement of Gases—Water Vapour and CO_2—Across Leaf Surfaces

During the day, when stomata are open, leaf water potentials may become very low. Values of -15 to -20 atm and lower are not unusual.

Table 2.7. Osmotic Pressure and Water Potentials of Leaves

Time of Day	*Andropogaon trifide Osmotic Pressure atm.*	*Ambrosia scoparius Water Potentials, atm*
Early morning	20-22	–7.10
Noon	22-24	–12.15
Mid-afternoon	25-27	–15.17

Clearly the largest free-energy changes occur between leaf and atmosphere.

Example 5. In a saturated atmosphere, is the decrease in water potential across the surface of an illuminated leaf sufficient to drive transpiration.

Let us grant that the sun raised the temperature of the leaf by only 1°, say from 25 to 26°. This would increase the vapour pressure of water in the leaf from 23.69 to 25.13 mm Hg. The $\Delta\psi$ for the movement of water from the heated interior of the leaf.

$$\Delta\psi = -\frac{RT}{V}\ln\frac{25.13}{23.69} = -80 \text{ atm}$$

is clearly greater than the $\Delta\psi$ for the movement of water into soil at the permanent wilting percentage (approx. –15 atm.)

We calculated earlier that the principal free-energy decrease in the transpiration stream occurs across the leaf surface. The opening and closing of the stomata are therefore likely to be major factors in the control of water movement. The part the stomata play is determined, of course, not by free-energy changes alone (that is, hydrostatics) but by the factors governing the rates of water movement (hydrodynamics). We can construct a hydrodynamic model of the leaf-stomata-atmosphere system from a consideration of the path of water movement.

The intercellular air spaces of leaves air spaces of leaves are in effective contact with most of the leaf parenchyma, including the palisade as well as the spongy parenchyma. The leaf parenchyma cell walls therefore present an evaporation surface to there numerous intercellular air space of the leaf. Water vapour can also diffuse through the epidermis and cuticle. In this segment of the transpiration steam, water movement is by diffusion, at least within and through the leaf. Fick's law (Equation 11) in a time-varying system is terribly messy.

Let us make the physiologically reasonable assumption of a steady state, with one concentration (activity or fugacity) of water in atmosphere outside the leaf (P_a) and another at the evaporating surface of the parenchyma cells (Pp). Now, instead of trying to calculate the tortuous path of diffusion of water vapour through the intercellular spaces, let us lump all the stomata and all the intercellular air spaces into two "apparent path lengths" of a diffusion system of unit are. Fick's equation (11) at steady state then reduces to Equation 27.

$$T = \frac{DA(p_a - p_p)}{L + S} \quad ...(27)$$

where T is the rate of transpiration, Pa and Pp are the partial pressures (or fugacities) of water in the atmosphere and at the evaporating surface, respectively, D is the diffusion constant for water in air, A is the cross-sectional area, and L and S are the apparent path lengths for diffusion in intercellular air spaces and through the stomata, respectively Equation 27 gives transpiration as a function of stomatal opening, but we must carry the analysis further before we can evaluate it. The apparent path length through the stomata, S, can be evaluated a consideration of stomatal geometry. The fluid path length S can be approximated by the relation

$$S = \frac{l / \pi ab + l / 2\sqrt{ab}}{n} \quad ...(28)$$

where l, a, and b are the depth, width, and length of the stomatal aperture, and n is the number of stomata per square centimeter.

Equation 27 can be inverted

$$\frac{1}{T} = \frac{L}{DA(p_a - p_p)} + \frac{S}{DA(p_a - p_p)} \quad ...(29)$$

which, for steady-state conditions, is a linear equation. L may now be evaluated by extrapolating 1/T for S = 0. Zelitch and Waggoner have shown that a linear relation obtains for tobacco leaves.

The reciprocal of transpiration 1/T, can be linear function of S. We took L, the apparent path length for diffusion in air, to be constant. Just outside the stomatal pore there will be a layer of quiet air, the barrier layer which represents part of the apparent path length L. The diffusion of water vapour from the pore through the barrier layer is commonly identified by the term "diffusion shells" in reference to the surfaces representing equal concentration of water vapour.

Brown and Escomb (1900) long ago demonstrated that the diffusion of gases through a system of pores through a membrane is remarkably efficient, up to 40 per cent as a great as diffusion from an open vessel of the same area: The thickness of the barrier layer is determined by the rate of movement of air outside the leaf. We can expect that at a sufficiently rapid rate of air movement, the component of L due to the barrier layer becomes negligible and L becomes equal to the diffusion path within the leaf. In fact with air turbulence, L virtually disappears.

The apparent diffusion path of water vapour in air is therefore due almost exclusively to diffusion through the barrier layer. Up to now we have treated the stomata kinetically, but what controls the opening and closing of these pores? There is a general agreement that the stomata are controlled by the turgor of the guard cells. This can be demonstrated by placing a section of leaf in a hypertonic sucrose solution medium. But in nature these turgor changes are triggered by a bewildering variety of factors.

If a generalization is possible we may say that the guard cells are programmed to close in response to conditions that either are unfavourable to photosynthesis or might be indicative of impaired photosynthesis: low light intensity, high internal CO_2 concentration, low water supply, and poor nutrition. The detailed effects of a number of these factors are reviewed by Health (1959). The mechanism of the control of turgor changes is a subject of intense dispute. Attention has focused on the chloroplasts of the guard cells: functional guard cells always contain chloroplasts; the action spectrum for stomatal opening corresponds to typical chloroplast absorption spectra; glycollate, a typical product of photosynthesis, is unusually effective in promoting opening, and inhibitors of glycollate catabolism are effective in inhibiting stomatal opening. If, as seem likely, the closing of stomata is caused by a passive loss of water from cells, and opening is active (that is, metabolically dependent), we should expect that low temperature and inhibitors should inhibit stomatal opening but no closing. Zelitch has shown that this is indeed so.

The immediate mechanism of opening is almost certainly osmotic, and the most likely immediate cause is a large increase in the potassium concentration in the guard cells. For the present one can only guess at the identity of the putative potassium pump and the means by which the various environmental factors control it. Substances that combine with sulphydryl groups are strikingly effective in inhibiting opening.

Metal-complexing agents also inhibit opening, but at certain concentrations, they inhibit closing as well. The fact that the phenylmercuri compounds when sprayed onto leaves penetrate only the epidermal layer provides an opportunity for a closer examination of stomatal physiology. After treatment with these substances, the stomata are "fixed" in various positions of closing for long periods of time with no direct effect on the parenchyma. Since different value of S can be obtained in this way, it becomes possible to test several deductions from the hydrodynamic model of gas movements through the stomata.

One inference is that, since the principal ΔG change is across the leaf surface, resistance to water movement is through the stomata. If this were so, the rate of transpiration would be dependent only on S (and L) and would be independent of soil water potentials. (In the absence of the inhibitor, water stress will influence stomatal opening.) Shimshi (1963) has shown that, contrary to expectation, soil moisture alters the I/T intercept (but not the slope) of I/Tvs. S plots. The there is a significant resistance to water movement within the plant itself. This additional resistance, which we can calculate from the X intercepts at different water potentials, may be in the parenchyma cell walls.

CO_2 Movements through the Stomata

We have seen how the movements of water vapour through the leaf can be dealt with as a simple kinetic system. The movement of another gas, CO_2, also crucial to the life of the plant, should follow virtually the same scientific model. We can modify Equation 29 in two ways and obtain a mathematical model which should predict the movement of CO_2. The diffusion constant for CO_2 in air is different from that of water vapour; so that D becomes D'. Also CO_2 must pass through the water layer surrounding the cell and diffuse to the chloroplasts, the presumed centers of CO_2 absorption. This requires another apparent diffusion path length, M.

The partial pressures Pa and Pp must be changed to correspond to the partial pressures of CO_2 in the air and at the chloroplasts, respectively.

Even though this distance M may be very small, the diffusion constant through water is enormously less (1/2,500) than that through air; consequently the M term may well be significant. The resulting equation (30) now expresses photosynthesis p as a function of M, L, and S. It is of course essential in using this equation that we employ conditions under which the rate of photosynthesis is limited by the CO_2 concentration and moreover that photosynthesis is a linear function of the CO_2 concentration.

$$\frac{1}{p} = \frac{M}{D'_w A(p_p - p_c)} + \frac{L+S}{D'_a A(p_a - p_e)} \quad ...(30)$$

If we eliminate L by the use of turbulent air, we can evaluate the M term. It is in fact a significant component of the resistance to CO_2 movement. One of the interesting conclusions from a comparison of the effects of stomatal aperture on transpiration and photo synthesis is that with moderate decrease in stomatal apertures, there is a disproportionately large decrease in transpiration. Much effort is now being spent in searching out the nature of stomatal control. Attempts are also being made to develop artificial cuticles with an eye toward rational measures of water conservation in arid land agriculture.

3

SOIL WATER

Water plays several different roles in the ecological relations of Plants and animals. As a material entering the organism water is important as necessary and abundant constituent of protoplasm, and the plant or animal body as a whole generally continues a large percentage of water—sometimes 90 per cent or more. Water is essential also as one material taking part in the photosynthetic reaction, through which energy becomes available either directly or indirectly to all living beings. Water is necessary as a solvent for food and as an agent for the chemical transformation of the materials within the body. Plant obtain mineral nutrients from the soil after they are in aqueous solution. Although most animals take in their food in solid form, this material must be dissolved before it can be absorbed by the blood and tissues. The intestines could be stuffed full of solid food and yet the animal would starve if no water were available for its digestion. Furthermore, water serves as a vehicle for transport or circulation within the bodies or organisms.

Minerals are carried up the stem of the plant by the transpiration stream. Water is the principal constituent of the circulatory, excretory, and reproductive fluids of animals, and it is necessary as a transfer agent at respiratory and olfactory surfaces. Water also acts as a regulator of temperature for plants and animals. For these essential purposes the proper concentration for water must be maintained inside the organism and, at the same time, the transfer of water must be suitably regulated. If the organism could live hermetically sealed in a capsule, the retention of the necessary amount of water would be easy, but exchange of this material between the organism and the outside world is necessary. The crucial need for a proper water balance is demonstrated by the consequences of even small water losses in some instances. During starvation, for example man may lose as much as 40 per cent of his body weight including half of the proteins and nearly all the glycogen and fat without serious danger, but if 10 per cent of the water content of the human body is lost, serious disorders result. If as much as 20 per cent of water is lost, death follows. The concentration of water divides the environment into aquatic and terrestrial habitats.

At first sight, one might say that more than enough water exists in the aquatic environment and that water is a problem only on land. Closer scrutiny shows, however, that water is tending to enter or to leave the organism too fast in almost every situation. In certain terrestrial habitats the water supply may be excessive for many organisms, as in some tropical rain forests where the air is often 100 per cent saturated and the ground is completely permeated. A visitor to such a habitat sees moisture condensing on every surface and hears the steady dripping of water from the vegetation. In the ocean in spite of the thousands of cubic miles of sea water a scarcity of water would nevertheless exist for many plants and animals! For these organisms the concentration of water is too low relative to the abundance of salt for proper osmotic equilibrium. Freshwater environments, viewed similarly in relation to osmotic balance, tend to contain too much water. This contrast between marine and inland aquatic situations is, it a sense, analogous to the difference between dry and humid climates on land. The maintenance of the proper water balance is thus a problem that must be considered in all habitats for all types of organisms.

Water Problem in the Aquatic Environment

Since the water in natural environments contains a varying amount of dissolved materials and usually a different amount from the fluids of the organism, the resulting differences in osmotic pressure raise a problem in regard to water exchange. In order to determine the tendency of water to enter or to leave the organism we must know something of the composition of the surrounding medium.

Composition of Natural Waters

Representative values for the amounts of the more numerous ions found in sea water and in the water of ponds and lakes are shown in Table 3.1. A larger amount of each of the common ions occurs in sea water than in typical fresh water. Hard fresh water contains more dissolved salts than soft fresh water, particularly with respect to the calcium and the carbonate ions. A comparison of the *relative* amounts of each type of ion in each kind of water may be made by recalculations in which sodium is given a value of 100 in each case (Table 3.2). In sea water chloride and sodium are the first and second most abundant ions, respectively whereas in typical hard fresh water carbonate is the most abundant with calcium second. In typical soft fresh water the calcium and carbonate ions are relatively less concentrated than sodium and chloride.

Table 3.1. Composition of some typical natural waters.

	Na	*K*	*Ca*	*Mg*	*Cl*	SO_4	CO_2	*Total (g/litre)*
Soft fresh water	0.016	—	0.010	—	0.019	0.007	0.012	0.065
Hard fresh water	0.021	0.016	0.065	0.014	0.041	0.025	0.119	0.30
Sea water	10.7	0.39	0.42	0.91	19.3	2.69	0.073	34.9

Table 3.2. Relative abundance of ions (by weight) in natural waters.

	Na	*K*	*Ca*	*Mg*	*Cl*	SO_4	CO_2
Soft fresh water	100	—	62.5	3.3	119.0	43.8	75.0
Hard fresh	100	76.0	310.0	66.8	195.0	119.0	567.0
Sea water	100	3.6	3.9	12.1	181.0	20.9	0.7

Samples of sea water from different localities are surprisingly constant in respect to the relative abundance of the major constituents. So much is this the case that for many purposes it is not necessary to measure the concentration of more than one ion. The standard procedure for determining the salinity of sea water is to find the amount of chloride ion present by titration and then to calculate the total salinity and the concentration of the other ions from the known ratio. The individual salts occurring in fresh water by contrast, vary widely—often with far-reaching consequences that will be considered later. At the moment our concern is with the maintenance of the proper osmotic balance between the inside and the outside of the organism.

Since osmotic pressure is determined by the total concentration of molecules and ions in solution, values for the aquatic medium will vary according to the total salt content of the water body (Table 3.3). Dissolved materials in soft fresh water vary in amount from practically zero to about 65 parts per million. Lakes and ponds with hard water exhibit a variety of higher values for salt content, but the value given in the table, 300 ppm, is representative. Salt takes may contain exceptionally large amounts of dissolved substances.

Table 3.3. Total salt contents of some typical natural water.

Soft fresh water	> 65.0 ppm	0.065 %	0.0065 %
Hard fresh water	300.0	0.300	0.03
Devils Lake, N.D.	11,000.0	11.0	1.1
Great Salt Lake	170,000.0	170.0	17.0
Baltic Sea		8.0	0.8
Ocean water		35.0	3.5
Red Sea		40.0	4.0

In contrast to the extreme variability in the salt content of inland waters, the open ocean tends to remain highly uniform in salinity over

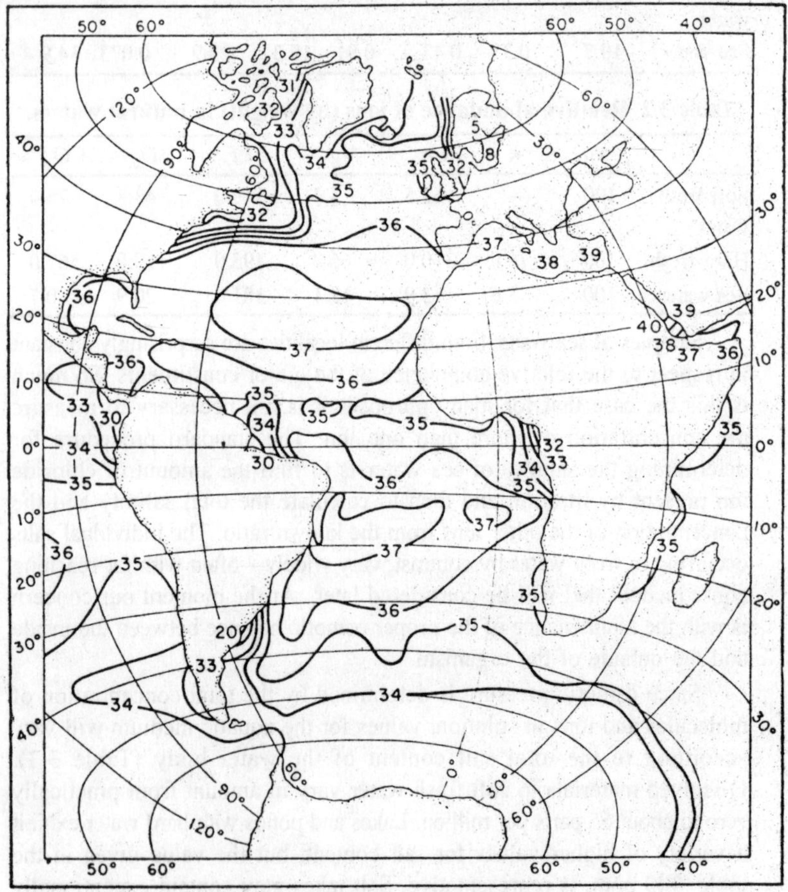

Fig. 3.1. The surface salinity of the Atlantic Ocean.

huge areas. A fish could swim up the middle of the Atlantic Ocean, for example, from the Cape of Good Hope almost to Iceland without encountering a salinity, lower than 35% or higher than 37% (parts per thousand). Ocean areas with unusually high evaporation, like the Mediterranean Sea and the Red Sea, exhibit salinities that run up to 40%. On the other hand, the salinity of the Baltic Sea is reduced to 8% or less due to the large inflow of fresh water. In river estuaries and in coastal areas receiving large amounts of run-off, salinity may vary over the whole range from nearly fresh water to full sea water. In smaller habitats, such as rock pools filled by ocean spray, the salinity may be typical of ordinary sea water on one occasion, may be greatly increased by evaporation a few days later, and may be reduced almost to zero by rainfall on another day. However for the marine environment as a whole, such situations are definitely exceptional. The great length and breadth of the ocean maintains a nearly constant salinity of about 35 to 36%.

The osmotic pressure of a molar solution of non-electrolyte is 22 atmospheres or 1700 cm of mercury. The freezing point of such a solution is depressed to—1.84°C, and this fact provides a convenient method for measuring osmotic pressure. In a solution of electrolytes with the same number of molecules the osmotic pressure would be higher because of

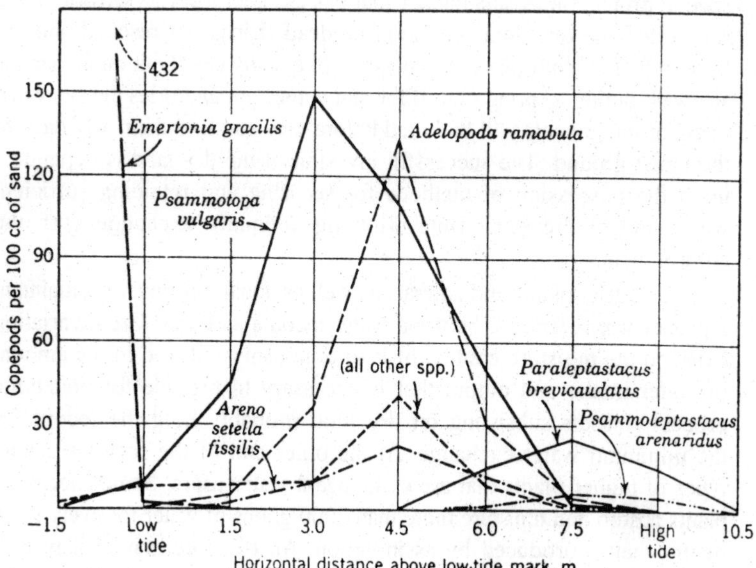

Fig. 3.2. Abundance and distribution of sand-dwelling copepods across the tidal zone at Nobska Beech, Woods Hole.

the dissociation. Sea water is a 0.55 molar solution, and, since the salts are dissociated, sea water has an osmotic pressure almost equal to that of a molar solution of a non-electrolyte. The salt content of the water in alkali soils may be even greater. Animals and plants living in sea water or in saline soils are therefore surrounded by liquids having a very high osmotic pressure.

Table 3.4 Typical values for osmatic pressure expressed as freezing Point Depression (=Δ).

	ΔWater	*Δ Invertebrates*	*Δ Teleost Fishes*
Mediterranean	2.14	2.2	0.8-1.0
Coastal Atlantic	1.79	1.8	0.7-0.8
Fresh water	0.03	0.1-0.8	0.5-0.6

Water Problem in the Terrestrial Environment

When we turn to dry land as an environment, the water problem becomes extremely acute. Everyone is familiar with the quick death which awaits aquatic organisms brought out on land. Many terrestrial forms can endure the land environment only if they remain in damp places. This is true for a great many species of plants and also for animals without special protection. A salamander that has escaped from its terrarium and run across the dry floor will soon be dead if it is not rescued and put back in humid atmosphere. Earthworms frequently crawl out on a concrete sidewalk during a spring rain. If the sun comes out and dries the sidewalk, many of the animals will be killed before they can find their way back to the moist ground. The successful invasion of the dry land is dependent upon the possession of methods for securing and retaining sufficient water and at the same time allowing adequate exchange with the environment.

Terrestrial plants and animals as well as aquatic forms must maintain a proper *water balance* between water income and loss. Great variation exists in the moisture content of land organisms and also in the amount of water intake and output that is necessary to provide for metabolic process. Insects subsisting on dry food probably hold the record for the minimum water exchange. At the other end of the scale are many types of higher plants that are extravagant water users. Some species of plants absorb and transpire more than 2000 grams of water for every gram of dry matter produced by assimilation. An oak tree may transpire as much as 570 liters of water in a single day. How many taxonomic groups have succeeded in colonizing dry land?

Fig. 3.3. Geographical distribution of mean annual rainfall.

Actually, only a very few of the prominent kinds of plants and animals have accomplished this. In the plant kingdom it is chiefly the vascular forms that are able to live in really dry places, but some low-growing bryophytes, algae, and fungi-particularly the rock lichens—form noteworthy additions to the list. Among the vertebrates, only the reptiles, birds, and mammals can live freely exposed to dry air. Although several phyla of invertebrates are reported in the land fauna, only the insects, spiders, and snails occur in important numbers in dry habitats. By contrast, there are a great many plant and animal groups that are exclusively aquatic.

Occurrence of Water in Land Environment

The only ultimate source of water for the terrestrial environment is condensation—chiefly in the form of rain. If precipitation were evenly distributed, it would cover the earth to a uniform depth of 1 m each year. Quite the contrary is the case. A glance will reveal the irregularity in the geographical distribution of rainfall. In the desert areas the rainfall is less than 25 cm per year and is inadequate for most organisms. In the black areas of the chart an excessive rainfall of 200 cm or more is reported each year. The mean annual rainfall on the south side of the Himalayas is recorded as 1232 cm. At the other extreme is Iquique, Chile, with an average precipitation of 0.125 cm per year.

Among the Olympic Mountains in the state of Washington the annual rainfall totals as much as 380 cm on the windward side of some ridges and as little as 25 cm on the leeward side. Seasonal variations in the distribution of rain may be of even greater importance than the geographical aspect. In many regions the division of the year into a rainy season and a dry season is of more ecological significance than the change in temperature between summer and winter. In the climatographs two situations are contrasted. In the Chicago region representing a temperate climate, an average an average of 5 to 9 cm of rain falls in every month of the year, but the average monthly temperature varies from about –4°C in the winter to about 22°C in the summer.

In the tropical climate of Barro Colorado Island the average monthly temperature changes by not more than 1 or 2°C throughout the year, whereas the precipitation drops to less than 3 cm per month in the dry season and rises to more than 40 cm per month in the wet season. Wherever the amount of precipitation varies greatly during the course of the year, the time of occurrence of rain in relation to the temperature cycle has great effect on the vegetation. Only those species will thrive whose varying needs for water in the different life stages are satisfied by the seasonal distribution of available moisture. In many of the grass-covered or forested areas of the temperate zone the major portion of rain occurs in the summer, but in other regions, of which southern California is an example, most of the precipitation occurs in the winter. In the latter situation we find a special type of vegetation with broad evergreen leaves known as the sclerophyllous forest.

In the prairie provinces of Canada the average annual precipitation of 50 cm would not be sufficient to support agriculture if the rain were evenly distributed through the year. However, the major portion of the rainfall occurs during the spring growing season, and also at a time before the rate of evaporation has become excessive, with the result that the available precipitation is used to best advantage. The existing combination of ecological conditions thus permits western Canada to be an important wheat-growing region. The foregoing discussion brings into relief the fact that in considering moisture conditions on land the actual amount of the precipitation is only part of the story. The other part, and

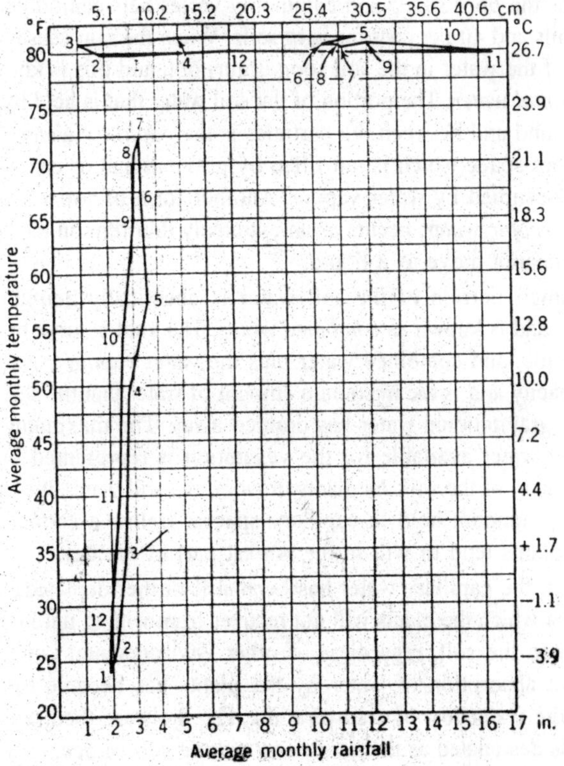

Fig. 3.4. Average rainfall and temperature for each month of the year as indicated by the numerals. Lower figure: temperate climate. Upper figure: tropical climate.

often the more important part, is the loss of water. Of the rain that falls upon the surface of the soil, part runs off immediately, part sinks in, and another portion is lost by evaporation. The amount of moisture at any one time and place depends upon the relative rates of the supply of water to the soil and evaporation from it—another example of an ecological factor whose value depends upon an equilibrium. In evaluating the water factor in the terrestrial environment both supply and loss processes must be taken into account.

Moisture in the Soil

Since the terrestrial environment is so varied, it is not surprising that the water factor is very complex, and the moisture in the soil and in the air will first be considered separately. When rain water enters the soil, it fills the spaces between the particles. The volume that can be filled is known as the *pore space* and commonly varies between 60 per cent for heavy soils and 40 per cent for light soils. When the rain causes, a certain portion of the water in the soil soon drains out, and this is known as the *gravitational water*. The portion of the soil water that is held by capillary force around and between the particles is the *capillary water*. That part of the pore space which is not filled by gravitational or capillary water may be occupied by water vapour. Another form of soil water, termed the *hygroscopic water*, occurs as an extremely thin film on the soil grains but this cannot move as a liquid.

A small portion of the soil water is chemically bound with soil materials and is known as *combined water*. The total amount of capillary, hygroscopic, and combined water plus the water vapour constitutes the field capacity and is the maximum amount of water that the soil can hold after the gravitational water has drained away. The maximum potential supply of water available for the vegetation is represented by the full field capacity of the soil. As plants draw on the soil water, they gradually reduce the amount held in capillary spaces. Below a certain moisture content plants tend to wilt in the hottest part of the day.

When the capillary water has been still further depleted, a point is reached at which the plants will not recover from wilting until more water is added to the soil, regardless of other environmental conditions. At this point absorption of water by the plants has become too slow to replace the water lost by transpiration. The moisture then remaining in the soil is designated as the *permanent wilting percentage*, or the *wilting coefficient*. Since little difference has been found in the abundance of soil water when wilting occurs for plants of various species growing in the same soil, the permanent wilting percentage is primarily a

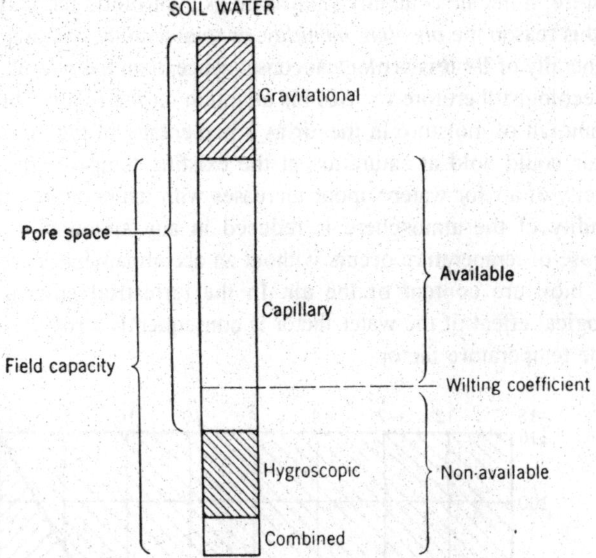

Fig. 3.5. Generalized diagram of the forms of soil water. The relative proportion and the value of the wilting coefficient differ according to the nature of the soil.

characteristic of the soil, and as such has great ecological significance. The amount of water between the full field capacity, as a maximum, and the wilting coefficient, as a minimum, represents the *available water*. A considerable amount of water may still remain in the soil after the wilting coefficient has been reached.

A fraction of the capillary water, the hygroscopic water, and the combined water, as well as the water vapour, cannot be obtained by the plant, and these constitute the *non-available water*. Soils vary considerably in the relative proportions of the different categories of soil water. In some instances the amount of non-available water may actually be larger than the available water. The fact that a portion of the water in the soil is available to land organisms and that water in the sea is unavailable to aquatic organisms not adapted to its osmotic pressure contributes further to the general concept of the universality of the water problem. From the ecological point of view the actual amount of water present in any habitat is not as important as its availability.

Moisture in the Air

For many terrestrial organism the moisture in the soil is chiefly important as constituting the principal source of water, whereas the moisture in the air is mainly significant as controlling the loss of water. For this reason the *absolute humidity*, or total amount of water in the air, is generally of far less ecological consequence, than the *relative humidity*. The ecologist therefore focuses his attention on the relative humidity or the amount of moisture in the air as a percentage of the amount which the air could hold at saturation at the existing temperature. Since the capacity of air for water vapour increases with temperature, the relative humidity of the atmosphere is reduced in any situation in which an increase of temperature occurs without an accompanying increase in the total moisture content of the air. In the terrestrial environment the ecological effect of the water factor is consequently strongly influenced by the temperature factor.

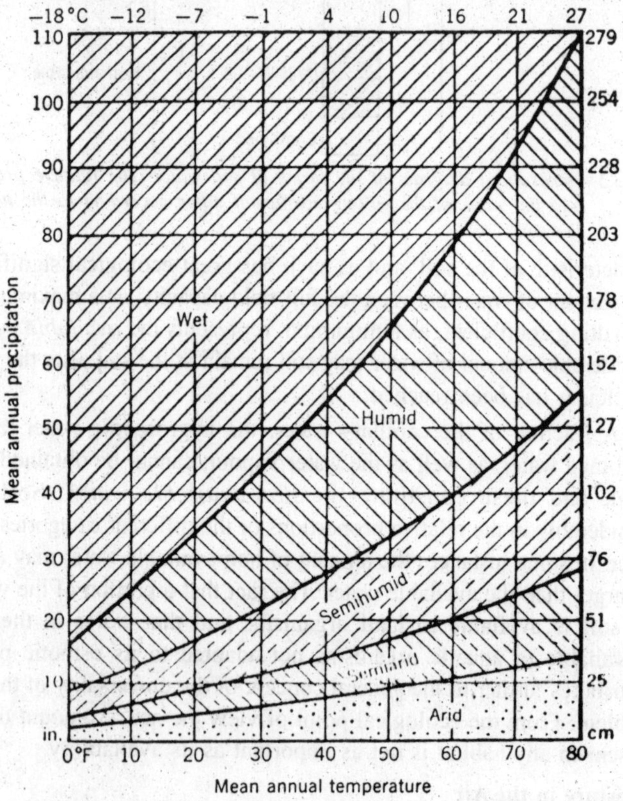

Fig. 3.6. Schematic representation of the influence of rainfall and temperature on climate.

Of two regions having the same rainfall is the drier in the ecological sense. The climate of a locality with a mean annual precipitation of 50 cm would be characterized as humid if the mean annual temperature were—7°C or less. On the other hand, another locality with the same rainfall would be regarded as having a semiarid climate if the annual temperature were 21°C or more. The contrast in climate between the Canadian prairie and the Mexican desert, both with about 50 cm of rain but with very different temperature conditions is an excellent illustration of this principle. The geographical variation in relative humidity is very great. Relative humidities of 80 to 100 per cent characterize the tropical rain forest. Regions reporting values of less than 50 per cent are regarded as having dry climates, and those with values of less than 20 per cent are extremely arid.

It is interest to note in passing that in cold winter weather the relative humidity inside our houses is rarely higher than 35 per cent. At any one locality the relative humidity may remain relatively constant for long period of time or may vary widely. On many oceanic islands the humidity is very nearly the same throughout the year. In other localities, characterized by wet and dry seasons, the humidity fluctuates widely from one part of the year to another. In certain situation as on the plains and in desert

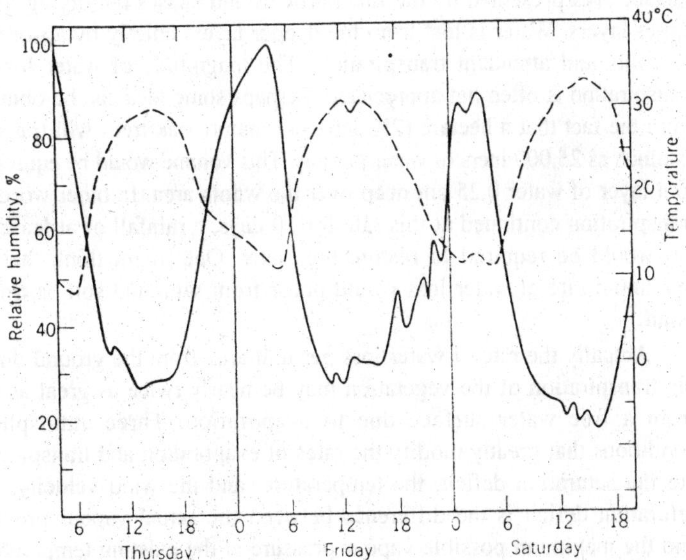

Fig. 3.7. Hygrothermograph record of temperature (broken line) and relative humidity (solid line) in the short-grass plains of central United States during days in early July.

regions considerable changes in moisture content may occur during the course of each day. Records made in a short-grass prairie in the United States showed a variation from a relative humidity of less than 30 per cent in the early afternoon to more than 95 per cent in the middle of the night. Obviously animals and plants living in such habitats must be equipped to withstand these rapid and extensive changes in the water factor, and their lives must be attuned to them. In deserts where daytime humidities are extremely low and evaporation is excessive, the greater relative humidity at night helps to relieve the critical moisture condition.

Many desert animals take advantage of this situation by going abroad only during hours of darkness; in some desert plants the stomata open only at night when transpiration loss is at a minimum. The foregoing has been a brief description of the amount of moisture in the air in terms of its relative humidity. As already implied, the chief ecological significance of the relative humidity is its effect on the rate of water loss. Terrestrial animals and plants lose water directly to the air by evaporation and transpiration. The water supply in their substratum is also reduced by direct evaporation from the soil and indirectly by the transpiration of the vegetation.

Evaporation takes place very rapidly from soil because of the great surface area presented by the fine particles and causes the drying of the upper layers. Water is lost from the deeper layers chiefly by absorption by roots and attendant transpiration. The magnitude of water loss by transpiration is often not appreciated. Perhaps some idea can be obtained from the fact that a hectare (2½ acres) of mature oak trees will transpire as much as 25,000 liters of water per day. This volume would be equivalent to a layer of water 0.25 cm deep over the whole area. In other words, if transpiration continued at this rate for 10 days, a rainfall of at lease 2.5 cm would be required to restore the water. One might think that the maximum rate of water loss would occur from saturated soil or from a pond.

Actually the rate of water loss per unit area from the ground due to the transpiration of the vegetation may be nearly twice as great as that from a free water surface due to evaporation. Three atmospheric conditions that greatly modify the rates of evaporation and transpiration are the saturation deficit, the temperature, and the wind velocity. The saturation deficit is the difference between the actual vapour pressure and the maximum possible vapour pressure at the existing temperature. The saturation deficit thus gives more information of ecological significance than the relative humidity alone since the saturation deficit provides a measure of the capacity of the air to take up additional moisture. An increase in the saturation deficit produces a rise in evaporation rate. An increase in temperature similarly speeds up the evaporation process.

A good indication of the magnitude of this influence is obtained from the amounts by which reservoirs are lowered in regions of different temperatures. During one year the water level in a reservoir in Ontario dropped 38 cm, a reservoir in California lost 2.4 m, and one in Egypt was lowered 3.6 m through evaporation alone. The distribution of plants often reveals the influence of temperature on direct and indirect water loss. On mountain slopes vegetation zones are found at different altitudes according to the exposure to the heat of the sun. Moisture-demanding species are restricted to higher levels on the side of the mountain toward the equator than they are on the cooler opposite side. The variation in the amount of rain required for a good growth of short grass furnishes another illustration of the effect of temperature. In Montana a precipitation of about 35 cm is sufficient, but in northwestern Texas a rainfall of 53 cm is needed to produce the same smooth amount of grass.

The explanation of this fact is easily apparent when it is realized that evaporation in the 6 summer months in Montana amounts to 84 cm, whereas in Texas it averages 137 cm. The wind velocity also exerts a major effect on the loss of water. With a gentle zephyr of only 8 mg per hr the transpiration of plants is increased 20 per cent over that which they exhibited in still air. A wind velocity of 16 km per hr increases the transpiration by 35 per cent and one of 24 km per hr increases it by 50 per cent. The desiccating action of warm dry winds sometimes prevents the invasion of windward mountain slopes by vegetation that grows perfectly well on the leeward sides of the same slopes where water loss is much reduced.

The harmful effect of excessive evaporation caused by the wind may also be seen in one's own garden. Here winter killing is occasionally due not to very low temperatures, but to a combination of wind and high temperatures! If a warm, wind blows for too long a period when the ground is frozen, the plants may lose water faster than their roots can obtain it and they will succumb as a result. Our discussion of moisture in the soil and in the air has indicated that living organism—particularly the higher plants—may exert a profound reciprocal influence on the water condition in the terrestrial environment. Water is a highly modifiable factor in land habitats, and living elements may act either to augment or to diminish the amount of moisture present.

Vegetation tends to each the rain, so that, if the precipitation comes in short showers, the rain may never reach the soil. The planet cover also tends to reduce the water content of the deeper layers of the soil as a consequence of its transpiration, but at the same time it adds moisture to the air. On the other hand, the presence of forest vegetation favours the reduction of evaporation at levels near the ground because it lowers the temperature, shows down the wind, and sometimes causes increased condensation. The amount of water present at any point in a land habitat and the rates gain and loss are thus seen to be the result of the equilibrium of processes both climatic and biological. The interrelations of the water factor and the vegetation are frequently very complicated and are the subject of elaborate studies beyond the scope of this book. In addition to presenting the general picture, the present discussion reveals the interdependency of vegetation and moisture as another outstanding example of the organism and the environment acting as a reciprocating system.

Microclimates

In the foregoing discussion of the variation in water loss under various conditions of humidity, temperature, and wind the values given were often those characteristic of extensive areas. Very frequently, however, the values of these factors in the immediate surroundings of the organism differ markedly from the regional values. In other words, the immediate climate of the organism is often sharply different from the average climate of the region as reported by standard meteorological records. The realization of this fact has led to the development by ecologists of the concept of the microclimate. In considering the effect of climatic influences on plants and animals it is absolutely essential to measure the environmental factors as they actually reach the organism.

The effective climate, upon which the life of the organism depends, is its microclimate. The characteristics of the microclimate may, or may not, be similar to the regional climate. Local variations in climatic factors of concern to many organisms may involve distances of several meters in the surroundings, but the microclimates of forms living in the soil, in rock crevices, under the bark of trees, or in similar confined habitats often have dimensions measured in fractions of a centimeter. The concept of microclimates is particularly important in relation to moisture because relative humidity varies widely within short distances in any irregular habitat.

In swampy woodland, for example, measurements have shown that a high rate of evaporation characteristically exist near the tops of the trees. At half this height the evaporation rate was reduced to 30 per cent in one instance. Lower down, near the damp soil, the animals and plants were subject to an evaporation rate of only 7 per cent of that near the treetops. We are consequently not surprised to find that this vertical change in moisture, as well as vertical gradients in light and in other factors, contributes toward the establishment of a characteristic vertical layering by species of the vegetation and sometimes also of the arboreal animal life in forests.

Among animals the difference in evaporation rate with distance above the ground is of particular importance to soft-bodies insects. Many of these are known to move higher in the vegetation at night, but during the day they are forced by excessive evaporative to retreat to lower levels. Contrasting values for the rate of water loss; in two microclimates were reported in an interesting experiment performed in the Middle West. Two sets of green ash seedlings were planted in pots; one set was placed in the midst of a sumac thicket, and the other set was put out on the open prairies only a short distance away. At the end of 8 days the first group of seedlings was found to have lost an average of 0.38 grams of water per sq cm of leaf surface.

The seedlings set out in the adjoining prairies had lost 0.88 grams of water per sq. cm, or more than twice as much during the same period. A traditional method of finding the approximate direction of north in the woods is by observing on which side of tree trunks the growth of green "moss" is thickest. In the North Temperature Zone in habitats in which the atmospheric moisture is near the critical point, algae and other non-vascular plants are characteristically more abundant on the north surfaces of tree trunks, because the higher evaporation due to in solution effectively retards growth on the south side. However, in many situations differences in exposure to wind may outweigh the effect of the sun in causing excessive water loss.

Thus, although the "north side" rule does not always hold, direct measurements show that the growth of epiphytic moes is a very sensitive indicator of microclimate, and the striking differences in the growth of algae on opposite sides of tree trunks can be seen by anyone who walks through the woods. An average value for the moisture condition of the forest would give no inkling of three small-scale ecological differences of crucial importance to the species concerned. The burrows of desert animals furnish another example of a microclimate in which the physical conditions are radically different from those of the region as a whole. When the kangaroo rat retreats within its burrow, it breathes air two to five times more humid than that outside.

Measurements indicate that, if this animal remained out of its burrow continuously, the rate of evaporation from its lungs would exceed the rate at which it could obtain water, and the resulting water, and the resulting water loss would eventually cause death. Temperatures in desert burrows have been found to be as low as 28°C at the same time values for the surface of the soil surpassed 71°C. Such critical differences in temperature and in other factors in microclimates will be considered further in subsequent chapters.

Meeting Water Problem on Land

Land plants can live only where sufficient water reaches them either through the air or through the soil. Some species can get along with relatively small amounts of water, and these plants, inhabiting deserts and other arid regions, are termed *xerophytes*. Plants eke out an existence on a reduced water supply in a variety of ways. Certain annual plants are adapted to maintain themselves in desert areas by remaining in the seed stage during the dry season and then passing rapidly through their life cycles when a rainy period occurs. Other desert plants, like the cacti, store water in succulent organs in sufficient quantities to tide them over long intervals of drought.

In addition, various species of non-succulent perennials, such as the creosote bush, sage brush, and many grasses are included among the xerophytes because they possess unusual ability to endure long periods of permanent wilting—in some instances running into years. Lichens and some mosses similarly can survive a surprising degree of drying out. Plants with water relations intermediate between those of the xerophytes and the hydrophytes are classified as *mesophytes*. Since land animals can move about, they can actively seek pools of water from which to drink or can obtain water in the form of snow or dew. Many are able to endure long periods between visits to the water supply.

Some terrestrial animals, such as frogs and toads, have the ability to absorb water through their skins from damp surroundings, whereas others find sufficient liquid water in the tissue fluids of their food. Another method of solving the problem of securing water in dry land habitats, and one employed widely by the insects, is the use of *metabolic water*, that is, the water resulting from the chemical breakdown of food materials. The clothes moth and the meal worm have the ability to obtain metabolic water in sufficient quantity to enable them to live their whole lives without any access to free water in the environment. Experiments indicate that the kangaroo rat of the Arizona desert may similarly secure the whole of its water supply as a metabolic by-product of its food.

Just as important as methods for obtaining water are means for resisting desiccation. Many of the same general types of adaptations having this effect have appeared in the course of evolution amongst both plants and animals. First and most obvious is the possession of a more or less impervious covering for the body. The cuticle of plants, the skin of birds and mammals, the exoskeleton of insects, and the mucous secretion of molluscs are examples. Another means for avoiding the excessive loss of water is the reduction of body surface. The leafless plants of the desert are an instance of this plan carried to an extreme. Other plants fold their leaves or turn them edgewise to the scorching heat of the sun. Many terrestrial invertebrates bury part or all of their bodies to reduce evaporation. It is well known that the termites which attack our houses are killed by exposure to dry air, but these pests can cross exposed foundations by constructing runways within which protection from excessive evaporation is afforded. In arid regions many of the smaller animals are active only during the night when drying is less intense.

Water may also be saved by reducing the loss in respiratory and excretory systems. Respiratory organs are often enclosed within a cavity as has already been mentioned in relation to the gills of land crabs. Lungs are commonly located well within the body with only a small opening to the exterior. The branching tubular tracheal system of the insects is another example, and in many forms the size of the external openings of the tracheae may be reduced by the spiracles. In a similar way evaporation from the internal cavities of the plant leaf is restricted by the guard cells of the stomata.

Reduction in the amount of liquid needed for excretion is a physiological adaptation that also aids the body in conserving water. Many animals with little available moisture in their habitats and in their food produce dry faeces. Furthermore, in the interest of saving water, it is desirable for land dwellers to dispose of their nitrogenous wastes with the minimum quantity of liquid urine. Mammals accomplish something in this direction by reabsorbing part of the water in the urine before it is excreted. The ultimate attainment in this matter is reached by those birds, reptiles, and insects that are able to excrete nitrogenous wastes as solid uric acid practically without water loss.

If lack of water becomes still more acute, it may be necessary for the organism to go into a dormant condition in which its need for water and its loss of water will be reduced. In the tropics many trees shed their leaves during the dry season. Other plants pass periods of drought as seeds or as spores. Many lower animals form cysts of one sort or another. Terrestrial molluscs close their shells with a thin epiphragm of secreted mucus, and more highly organized animals, such as the ground squirrels, go into a state of torpor in their burrows. Dormancy of this sort occurring under conditions of heat and drought is referred to as *aestivation*, and is found characteristically in desert and tropical communities. Using various combinations of the methods for procuring and conserving water discussed above, plants and animals in arid habitats may live for months or years without rain or access to free water.

Influence of Moisture on Growth and Distribution

The growth of plants is often more directly and obviously affected by the water factor on land than is that of animals. The availability of moisture influences not only the rate of plant growth but also the growth form. Since the roots of most land species will not grow into saturated soil nor into soil devoid of available water, plants will be shallow rooted either if the water table is high or, in the contrasting situation, if the soil water is limited to the uppermost layer. In the latter case growth will be confined to the rainy season. The growth of plants whose roots extend to the permanently moist subsoil are largely independent of rain periodicity.

Competition for water and the horizontal spread of roots often control the spacing of plants in regions where water shortage is critical. The roots are seen to be spread widely just beneath the surface of the soil where they can absorb whatever rain falls before it evaporates. This cactus will capture all the available moisture within the area shown and prevent any other perennial plant from gaining foothold in close proximity. Desert vegetation is often spaced out in a strikingly regular pattern as a result of this intense root warfare for water.

The water factor also controls the geographical distribution of plants on both small and large scales. Such trees as willows and cottonwoods are characteristically limited to the moist banks of water courses because their seeds will not survive more than a few days unless the soil upon which they fall is wet. The seeds of other species can remain dormant in a completely dry condition for months or years but will germinate only when sufficient water is available in the soil immediately surrounding them. In other instances the critical period for the survival of a plant in relation to its water need may occur, during adult life. Plants differ greatly in the amounts of water that they must absorb from the soil and transpire. Those with a high water requirement, or *transpiration ratio*, are limited to habitats where the supply of moisture is adequate.

Differential ability to withstand drought will also play its role in controlling geographical distribution. If the soil moisture drops below the permanent wilting coefficient, most annuals and other tender plants will die within a matter of hours or days unless the soil water is renewed. On the other hand, succulents and plants that can aestivate are able to with stand varying periods of drought extending into years with some species. Thus the duration of drought periods is of crucial importance in determining whether a region may be inhabited by different plant types. Local differences in the moisture factor were clearly shown to influence the species composition of the forest vegetation in a study made in Indiana.

A series of observations in a line running from north to south over a ridge revealed the fact that soil moisture was 20 per cent higher on the north side of the ridge than on the south side. Moreover, the rate of evaporation was more than twice as great at stations on the south slope than at those on the north slope. Quadrat counts of the number of trees on the two sides of the ridge gave the tabulated results. Although other ecological factors undoubtedly played some part, moisture conditions were chiefly responsible for the very pronounced differences in the occurrence of the various species on the two sides of the ridge.

Species	*North Slope*	*South Slope*
Maple	202	31
Beech	112	0
White oak	0	80
Black oak	0	70
Hickory	4	44

Another example of the control of distribution by the moisture factor, and one on a much larger scale, is found in the regional occurrence of the natural vegetation in the United States. A glance will reveal the fact that in the central part of the country, leaving out of account the irregularities due to the mountains, the zones of principal vegetation types tend to run north and south. In the eastern part of the continent the typical natural vegetation is forest (in areas where man has not interfered). West of the forest zone is a belt of tall-grass prairie, and beyond that, running from Montana and North Dakota down to Texas is a zone of short-grass rangeland. Still farther westward lies the desert belt. This zonation is obviously not controlled primarily by temperature since the temperature belts are seen to run generally east and west. If we again leave out of account the mountain and coastal regions. Similarly, no good correlation will be found between the occurrence of these principal zones of vegetation and the annual precipitation taken by itself.

As already explained, regional differences in evaporation must also be considered, and water loss in turn depends in part on temperature. Both the supply and the loss aspects of moisture conditions may be taken into account by calculating-precipitation evaporation ratios for the "frostless season," i.e., the growing season for plants between the last spring frost and the first autumn frost. The pattern of the distribution of these ratios in the United States shows a general agreement with the occurrence of the major vegetation zones.

The natural forest vegetation of the eastern part of the country occurs where the precipitation-evaporation radios are high. In the Southeast and Northeast, the P-E ratio exceeds 100 per cent; that is, more rain falls during the course of the average frostless season than evaporates. Incidentally, this fact accounts in part for the excessive amount of soil leaching in these regions. In the Middle West precipitation-evaporation ratios between 80 and 100 per cent exist. Values below 100 per cent indicate that the potential rate of water loss is greater than the rate of water supply. The line representing a P-E ratio of 60 per cent runs almost due south from Dakotas to the southernmost point of Texas.

It will be noted that this line agrees closely with the division between the tall-grass prairies and the short-grass rangelands. The 60 per cent line also approximates the position of the division between the pedocals and the pedalfers discussed earlier. P-E, ratios of less than 20 per cent are found in the southwestern states, and their occurrence corresponds to the distributions of the desert vegetation. In view of the modifying influence on plant distribution of such factors as soil type and temperature and of the interdependence of these factors with moisture it is clear that no simple relation between the type of vegetation and any single physical influence can be expected. However, the broad correspondence between the zones of vegetation and the availability of moisture will suffice to emphasize the major role of the water factor in the region under discussion in controlling both the soil and the plant life that develops with it. The application of the P-E ratios to those distributional problems has again brought into relief the fact that the balance between supply and loss is the most crucial aspect of the moisture factor for the vegetation.

Soil Water Potential

Water in soil may exist in three states viz., solid (ice), liquid (soil solution) and gas (vapour). The water that enters into the soil either by rain or irrigation may pass through the surface soil layers to the water table and is known as 'ground water'. The water which is not drained deep into the soil profile is either retained in the soil pores and channels or on the surface of the soil particles. These films of water are bound to soil particles by strong molecular forces of adhesion and cohesion. As soil dries out, these water films become thinner and thinner and more tightly bound. This firmly held water was called earlier as firmly bound water. These water molecules which cannot be separated even by centrifugal action develop forces several tens of thousand times that of gravity. This layer of firmly bound water has a thickness of the order of few tens of molecular coats and in its turn, links and orients water molecules by which it is surrounded.

The latter form of water was earlier known as loosely bound water. It is, however, not possible to draw a boundary line between firmly bound and loosely held water. The distribution of water in the soil mass depends on the nature of the solid phase, i.e. the composition, the size distribution and the arrangement of soil particles. The hydrophilic nature of soil particles also determines the thickness of the water film and the force of retention on the solid surface. The large specific surface, both internal as well as external, of the colloidal soil particles is available for water absorption. In view of the complex nature of soil particles, their surfaces and pore-size distribution, several other types of water have been recognized in soil, viz. Vapour, hygroscopic, pellicular, capillary, gravitational, water in solid state, quasicrystalline and chemically bound water. The water vapour fills the pore spaces which are not filled with liquid water.

The hygroscopic water is the water that condenses on the soil particles. The water is absorbed when dry soil particles come in contact with humid air. The pellicular water forms on the particles under the influence of one molecular forces of adhesion. Capillary water is the water held by the forces arising due to presence of thin bore capillary size pores and the air-water interfaces. Gravitational water is the free water that is easily drained by the force of gravity. The chemically combined water is in the crystal structure of clay minerals which cannot be removed even by drying the soil at 110°C. The amount of water present in a soil at a given time can be expressed quantitatively.

The water content of a sample of soil is defined as the amount of water lost when dried to constant weight at 105°C. The water lost is expressed either as mass of water per unit mass of soil or volume of water per unit volume of soil. Recent studies have established some functional relationships between the soil water content and the different forces with which water is held by the soil or the energy status of water in the soil. Because different terms used earlier in classifying soil water are too arbitrary to describe the state of soil water quantitatively, the energy concept of soil water is described using two fundamental approaches, one from the mechanical point of view based on the potential theory of classical physics where the concept of soil water potential will be discussed in detail and the second from the thermodynamical point of view where the chemical potential of soil water will be discussed. In the first approach the forces are considered with respect to mechanical equilibrium assuming that the soil-water system is under isothermal condition and there is no significant movement of ion species. The thermodynamic approach, however, accounts for these effects.

Potential Concept

The most important concept in nature is the concept of energy because it is a fundamental entity common to all forms of matter in our physical world. Water retention and movement in soil is a consequence of energy effects. Commonly speaking the capacity of doing work is called 'energy'. Thus, closely associated with energy is the concept of work. In a physical sense, the work is done only if a force applied on an object causes its displacement, i.e. the work is done when an object is moved by a force which has a component acting in the direction of motion. Quantitatively, in its simple mechanical form, work is defined as the force times the distance through which the force acts, i.e.

$$W \text{ (work)} = F \text{ (Force)} \times h \text{ (Distance)}$$

Consider an example of the work done in lifting the mass, *m*, of water to a height, *h*, from *b* to *a* above the water level (reference level). It will be necessary at this point to define the force. According to Newton's second law of motion, when a body is acted upon by a constant force, its resultant acceleration is in the direction of the applied force and is given by

$$a = \frac{F}{M}$$

or $$F \text{ (Force} = m \text{ (Mass)} \times a \text{ (Acceleration)}$$

Therefore, by Newton's second law of motion, the force required to lift the mass, *m*, is equal to its own weight.

$$F = m.g$$

where g is the acceleration of gravity. Substituting the weight . mg for F in Eq.(1)

$$W=mgh$$

If a 5 *g* mass of water is raised vertically to a distance of 1 m, the work done will be

$$W \times 5 \times 980 \times 100 = 4{,}90{,}000 \text{ dynes-cm or ergs.}$$

We may state here that any type of potential should have similar definitions or expressions as written above. We may show this fact by defining the more familiar term the 'electrical potential' that is related with the transfer of charge. In physics the electrical potential, V_e, of a charge *e* placed in the field is defined as the amount of work done or potential energy stored, per unit positive charge in carrying any charge *e* from the reference point to the point in question, *i.e.*

$$V_e = \frac{E_p}{e}$$

where V_e = electrical potential, volt; and *e*=charge moved into the field, coulomb (1 coulomb =3 × 10^9 electrostatic units of charge).

In MKS system the unit of electrical potential is volt. If two points differ in potential by 1 volt, 1 joule of work is required to move 1 coulomb of charge from one point to another,

$$1 \text{ Volt} = \frac{\text{Joule}}{\text{Coulomb}}$$

Soil Water Potential

Because the velocity of flow of water in soil is often very low its E_k is negligible and its potential energy, E_p is primarily responsible for determining the state and movement of water in soil, The potential energy of soil water varies considerably from point to point. This difference of potential energy between two points gives rise to the tendency for water to flow. The water flows from the point where the potential energy is higher to where it is lower in the direction of greatest decrease in potential energy. The rate of decrease of E_p with distance is the moving force causing flow.

The potential energy is acquired by the water whether the work is done against gravitational, elastic, osmotic, electric or magnetic forces. The term soil water potential denotes the specific potential energy of soil water to that of water under standard reference state. Before we discuss the potential energy of water, we may first experience the major forces that act on water in the soil. They are:

1. Adsorptive forces-long range electrical forces and the short range London van der Waal's forces.

2. Osmotic forces related with solutes.
3. Gravity, inertial or body forces,
4. Hydrostatic or forces associated with pressure gradients.

If temperature of the system is not constant, an additional force field arising due to temperature will also be acting. For simplicity, we assume that the soil system in question is under isothermal conditions. For any force acting on soil water, a corresponding potential component can be specified. For gravity there will be a gravitational potential, ϕ. Similarly, other potentials that are associated with above mentioned forces will be pressure potential, ψ, osmatic potential, ω, adsorptive potential, α, etc. Since potential is a scalar quantity, it can be expressed as a algebraic sum. The total soil water potential, Φ, can thus be expressed as summation of different potential components:

$$\Phi = \sum_k \Phi_k$$

or
$$\Phi = \phi + \psi + \alpha + \omega + ...,$$

where Φ = total soil water potential, erg g^{-1}; ϕ=gravitational potential, erg g^{-1}; ψ = pressure potential, erg g^{-1}; α = adsorptive potential, erg g^{-1}; and ω = osmatic potential, erg g^{-1}. The total soil water potential can thus be defined as 1 "the amount of work that must be done per unit quantity of pure water in order to transport reversibly and isothermally an infinitesimal quantity of water from a pool of pure water, at a specified elevation and at atmospheric pressure, to the soil water at the point under consideration."

Reference State

The selection of reference level in all the above examples is purely arbitrary. In any soil water system the zero energy level in known as reference state. The conventional characteristic of reference state of water are the following:

1. Zero hydrostatic pressure: The reference level is considered to be free flat surface with zero hydrostatic pressure.
2. Zero elevation: All points which are measured in the soil water system, should be from this reference level.
3. Atmospheric pressure: Generally, one can use barometric pressure but it continues to change depending upon weather conditions. Therefore, standard pressure of standard atmospheric pressure is used.

   1 atmosphere = 1.013×10^6 dynes cm^{-2} = Pressure exerted by a column of water of height h = 1035 cm (with ρ_w = 1 g cm^{-3} and g = 980 cm sec^{-2}).

In metric system, the unit of pressure in atmosphere is replaced by bar(= 10^6 dynes cm^{-2}).

4. *Pure water*: The water at reference state has no dissolved gases or solids and free of any other impurities.

Water Movement in Soil

When water, whose source; is either rainfall, or irrigation, is applied to soil, it enters the soil pores replacing the air contained in them. If sufficient quantity of water is available, the entire pore space may be filled with water and the excess water would move downward by a physical process known as saturated flow. The flow of water in the upper saturated soil layers could be compared to the flow of water through the pipes. Such a type of flow results primarily from gravitational and pressure potential gradients and is comparatively rapid. If, however, the application of water is limited—as happens when rain is scanty, or irrigation water is in short supply—the rate of such movement is drastically curtailed.

If evaporation is taking place at the soil surface and the plants are absorbing water from wetted soil zones, the soil water from the profile is rapidly depleted. Under such conditions, the soil pores will be only partially filled with water and the soil water potential decreases further. The flow of water under these conditions will occur in response to the differences in soil water potentials in the different regions of the soil profile. Water moves from a region of high potential to depleted soil water zones (low potential). Such a type of water movement is termed as unsaturated flow. Although movement of water in soil is mainly in liquid phase, its movement in vapour phase; also occurs due to vapour pressure gradients caused by temperature and salt concentration gradients.

The movement of water in the vapour state may take place within or over the soil surface. From the soil surface, the water is lost to the atmosphere by diffusion and convection. This phenomenon is known as evaporation Due to differences in the vapour pressure, the water vapour in the soil diffuses from one region to another. Thus, water movement in soil may be of three types: (1) saturated flow; (2) unsaturated flow; and (3) vapour flow. Under field conditions, the movement of water is complex. During irrigation and subsequent redistribution of water in the soil profile and its loss through evaporation and deep drainage, one may encounter all the kinds of water movement phenomena described above. The nature of the soil profile, its texture, structure and pore size distribution determine to a large extent how water penetrates and is retained in the soil.

In a uniformly moist homogenous soil, water moves at nearly the same rate in all the directions. Water introduced at the centre of a vertical soil column is observed to move almost as far upwards as downwards and laterally. In this case, the gravitational potential gradient is negligibly small in comparison with the soil suction gradient. This circular pattern of water is also observed in furrow irrigation. When the water moving downward through a well-aggregated soil encounters a clay layer this layer is wetted readily because of the lower suction in the wetting front in the coarse soil. However, the downward movement is appreciably slowed down due to greater resistance to water in passing through fine pores in the clay layer. Water tables generally build up over the layer of clay. On the other hand, when the water moving downward through the fine textured soil encounters a layer of course sand the downward movement is temporarily stopped until the soil suction in the wetting front is sufficiently reduced.

This is due to the presence of large macropores in the course sand which are too large to take in water from the macropores of the soil. However, when the soil above the course sand is almost saturated, the water moves into and through such sand layers rapidly. Though the soil above the course sand gets very wet, water tables are not built up because the underlying coarse material will conduct away all excess water rapidly, acting as a drain for the overlying soil. Similar is the case with respect to the soil overlying gravels. Therefore, soil overlying sand and gravel absorb 2-3 times as much water as the same soil material if it did not overlie sands and or gravel. In the presence of sand lens, greater water movement is observed where the sand lens is in direct contact with free water supplied from the top. The aggregates also behave in the same way as the sand layer. However, the amount of water moving through the aggregates is limited by a relatively small number of contacts between the aggregates. It can be seen that the presence of course soil aggregates considerably retards the downward movement of water in soil.

Similarly, coarse soil, which infiltrates water more rapidly than fine soil, transmits water more slowly than fine soil during the later stages of drainage due to the higher conductivity of fine soil relative to the coarse soil under unsaturated conditions. Incorporation of crop residue creates favourable porosity conditions for better water penetration. In vertical mulching if the much and channel is open to the surface for free entry of water, such a channel would help in allowing the water to enter deep into the soil, thereby reducing the surface runoff. If the channel is not open to the surface such a phenomenon is not observed. This also shows that favourable surface conditions help water entry while under unfavourable surface conditions, even moderate rainfall does not infiltrate and causes runoff and soil erosion. If the flow of water in a saturated soil can be compared with the flow of water through tubes or pipes.

Poiseuille's equation can be used to quantify it. The soil pores, however, do not resemble uniform smooth pipes or tubes. In fact, soil pores are highly irregular and tortuous. Therefore, Poiseuille's may be of little help in describing the nature of water flow in such a complex porous medium. The numerous constrictions or necks in the pore space and dead-ends restrict the flow of water. As a result, the flow velocity varies considerably from point to point. However, if the soil pore space can be imagined as a bundle of uniform capillaries of radius r and with their surface area in the direction of the potential gradient, we may still apply Poiseuille's equation and can expect better approximation of the discharge through soil. To visualize this, consider a cylindrical soil mass, of macroscopic radius a containing *n* number of pores of radius *r*. The overall flow rate will be equal to the sum of the separate flow rates through the individual tubes.

Darcy's Law

The flow of water through porous materials such as sand was experimentally studied, more than a century ago (in 1856), by a French engineer Henry Darcy. He investigated the behaviour of vertical sand filter beds for the water supply to the town of Dijon in France. He measured the quantity of water filtering through the sand bed and showed that the volume of water passing in unit time depended upon several factors, namely, the cross-sectional area of the bed, the thickness of the bed, the height of the water on the filter bed and the proportionality constant, K. The value of K as observed by him depended upon the nature of the bed. The famous experimental Darcy formula is

$$Q = K\frac{A(h+L)t}{L}$$

where Q=volume of water passing through sand filter, cm^3; L=length of the sand filter bed, cm; A=cross-sectional area of the sand filter bed, cm^2; h=height of water on the filter bed, cm; t=time, sec; and K=proportionality constant cm sec^{-1}.

Substituting q = Q/At, Eqn. can be rewritten as

$$q = K\frac{(h+L)}{L}$$

where q = volume of water passing through unit cross-sectional area of material in unit time and is termed as flux, cm^3 cm^{-2} sec^{-1} or cm sec^{-1}; h +

L = hydraulic head difference between the inlet and outlet ends of the coloumn, cm; and (h + L)/L = hydraulic gradient, dimensionless.

From the foregoing discussion, it is clear that the discharge rate (the volume of water flowing through the soil column in unit time t) is directly proportional to the cross-sectional area, A, and to the hydraulic head difference and inversely proportional to the length of the soil column, L. The proportionality constant, K, is known as hydraulic conductivity. Hydraulic conductivity is thus the constant of proportionality in Darcy's law as applied to the viscous flow of water in soil. It is the flux of water per unit gradient of hydraulic potential.

FLOW OF WATER IN UNSATURATED SOIL

We have so far considered the flow of water in a soil where sufficient water was available to fill the pores and the movement of water was in accordance with Darcy's law. We may now apply Darcy's law to a soil where the pore space is not completely filled with water, that is, the effectively conducting pore space is not the total pore space but a partially saturated one. The flow of water under such conditions is known as unsaturated flow. A porous medium may be unsaturated in several ways. It may contain two fluids—a liquid and a gas—or two liquids and a gas. The saturation can be defined with respect to one of the fluids.

In case of soils, we generally consider two fluids, namely, water and air. Therefore, here, the degree of saturation refers to the portion of pore volume filled with water. Thus an unsaturated soil consists of three phases : a solid phase, a solution phase and a gaseous phase. In unsaturated soil the fluid may form a continuous phase or be separated by soil particles. We shal! consider both air and water forming continuous phases. It is, however, possible for the air bubbles to be entrapped among the soil particles. Let us consider a flow system similar to Darcy's filter bed. In the system described in this case, the pores are only partially filled with water, that is, the soil column is unsaturated and water flows under the influence of negative hydrostatic pressure head or suction head difference. In such a situation, the hydraulic head at the inflow (A) and outflow (B) boundaries of the soil column will be as:

Hydraulic head, H	*Pressure head, h*	*Gravitational head, z*
H at A =	$-h_1$	$+z_1$
H at B =	$-h_2$	$+z_2$
ΔH =	$-(h_1 - h_2)$	$+(z_1 - z_2)$

Also, $\Delta H = H_1 - H_2$.

Here, h_1 and h_2 are the pressure heads with a negative sign and z_1 and z_2 are the gravitational heads.

If h_1 and h_2 are maintained constant and the column length is small, the flow will soon attain a steady value. According to Darcy's Law, the flux through the column will be given by

$$q = K\frac{\Delta H}{L}$$

Since the soil column is not uniformly saturated, the gradient along the column may not be constant. In a long soil column, the suction will vary along the axis of the column and so will the water content and the hydraulic conductivity. If the steady flow velocity is maintained everywhere in the column, the gradient of potential should vary along the axis of the sample; being high where the conductivity is low and *vice versa.* With the increase in suction, the suction gradient will also increase and the conductivity will decrease. However, in the case of is stratified soil column, consisting of coarse sand overlying fine sand, with a steady downward flow, the distribution of soil suction with depth shows abrupt discontinuity at the junction between the two layers. The soil water suction in the fine sand which is in contact with the water table depends upon the amount of water that has entered it. Since the downward movement of water in fine sand will be appreciably reduced at the junction between the two due to the presence of fine pores, which offer a high resistance to water passage, the soil suction in coarse sand remains low compared to fine sand showing a high gradient at the junction. When steady flow is approached a linear distribution of suction is observed in both the layers and the flux is equal to the effective hydraulic conductivity of the column which is determined by the unsaturated transition zone. If, however, the fine sand overlies a coarse sand the picture is different. In this case, suction increases at the junction and the zone above the junction has a higher suction in the fine sand than the limiting case.

The water will enter the course soil only when the suction is reduced to the value determined by the pores in the coarse soil at the junction. A cultivated soil consists of aggregate containing discrete groups of smaller pores within the aggregates (also called crumb pores or macropores) separated by a more continuous system of large pores between them, that is, the intercrumb pores or macropores. A well-aggregated field soil which has drained to attain a uniform moisture content after saturation can be compared to a bed of aggregates whose intercrumb pores have drained to a uniform water content, leaving the crumb pores saturated. Such a pore system with a bimodel pore size distribution is characterized by a double sigmoid water characteristic curve. In all the pores—crumb pores and intercrumb pores-are saturated and water flow takes place in the larger macropore channels, although water can also move through the aggregates. It is the size and distribution of these macropore channels which determine the hydraulic conductivity of the soil.

With a small suction, the macropores will drain leaving the micropore region still saturated. The liquid continuity in the macropores is broken and the water moves through the aggregates which form the continuous path for water movement by means of lenses of water near their points of contact. The hydraulic conductivity of the aggregates decreases rapidly as the area of connecting lenses becomes smaller with decrease in water content. With further increase in suction, the macropores themselves begin to get desaturated, causing further reduction in hydraulic conductivity. The experimental results agree with the above reasoning. It appears that a well aggregated surface soil could be helpful in soil water conservation. The hydraulic conductivity of the bed of aggregates may be very low even if the crumb pores are nearly saturated. Therefore, during periods of evaporation from the soil surface a mulching effect is obtained. It should, however, be noted that a well-aggregated will may not attain complete saturation in practice because of entrapment of air inside the aggregates.

During heavy rainfall or flood irrigation, water in the macropore channels may be saturated, leaving the centre of the micropore region filled with air. During the uptake of water by soil crumbs, the entrapped air may be compressed to such an extent that the uptake may practically stop. There is also the possibility of the explosion of aggregates and the breakdown of soil structure when the pressure inside the crumb exceeds the breaking point. A number of experiments have been conducted to determine the variation of hydraulic conductivity with soil water content or suction. It is seen that the hydraulic conductivity is sharply reduced with an increase in soil suction. The shape of the curve for the three soil textures is different.

The saturated hydraulic conductivity of sand is the highest and that of clay is the lowest. However, with an increase in soil suction, the trends are reversed. The unsaturated hydraulic conductivity of sand decreases more rapidly with the increase in suction while that of clay decreases less sharply. At a higher suction value, the conductivity of sand is much lower than that of clay soil. In general, the hydraulic conductivity is very sensitive to water content or suction and ranges over a factor of 10^5 for the water contents measured in the field. The foregoing results indicate that under unsaturated conditions, the flux is not linearly related to the hydraulic gradient as in the case of saturated low.

Darcy's equation is thus non-linear under unsaturated conditions, in the sense that the hydraulic conductivity depends on soil water content, θ, or soil suction, τ. Since K is independent of flux and/or suction gradient, a plot of flux versus hydraulic gradient at a fixed average water content θ will give a straight line passing through the origin. A family of such straight lines can be obtained from careful experimentation. The slope of the line is the hydraulic conductivity at the indicated water content. Since the hydraulic conductivity is no more constant but highly dependent upon soil water content or suction, attempts have been made to modify Darcy's law for unsaturated soil by introducing functional relationships between hydraulic conductivity and water content or matric suction. For unsaturated soils Darcy's law has been written as:

$$q = K(\theta)i$$

where i = hydraulic gradient.

In the vector form, Equation can be rewritten as

$$q = -K(\theta)\,\nabla H$$

or

$$q = -K(\tau)\,\nabla H$$

These equations show that K is not constant but changes with water content or matric suction. If the soil water contents is known at a particular point, the flux at that point will be [from Equation].

q	=	K	×	i
Flux at a point		Hydraulic conductivity at that point		Hydraulic gradient at that point

No universal relationships are available for hydraulic conductivity and soil water content or suction. Nevertheless, several empirical relationships have been proposed. The following equations have been found useful in practice as it is difficult to predict precisely the unsaturated hydraulic conductivity from fundamental soil properties:

$K = a\tau^{-m}$

$K = a(b + \tau^m)^{-1}$

$K = K_s\,[1 + (\tau/\tau_c)^m]^{-1}$

$K = K_s\,(b + \tau^m)^{-1}$

$K = K_s\,(1 + c\tau^m)^{-1}$

$K = a\theta^m$

$K = K_sS^m$

where K = hydraulic conductivity at any soil water content; K_s = hydraulic conductivity of the same soil at saturation; a, b, c and m = empirical constants characterizing the soil. The value of m varies from 1-2 (for fine textured soils) to 4 or more for coarse textured (sandy) soils; τ = matric suction cm; τ_c= metric suction at which K=½K_s : θ = volumetric water content, cm^3 cm^{-3}; and S=degree of saturation.

Equation is not valid when the matric suction approaches zero. Since the water content at a given suction depends upon whether the soil has been undergoing drying or wetting (hysteresis) the relation between conductivity and suction will also depend upon hysteresis.

INFILTRATION

In water management and conservation studies, accurate information on the rate at which different soils will take in water under different field condition is required. The rate of water entry into soil varies widely between different soil types and also within a single soil type, depending upon soil water content and management practices. When water is applied to the soil surface through irrigation or rain, it subsequently enters the soil. If the supply rate is more than the rate entry into the soil, the excess will accumulate over the surface and may flow as runoff. The time rate at which water will percolate into the soil through its soil-atmosphere interface is known as infiltration.

Quantitatively infiltration rate is the volume of water entering into the soil per unit area in unit time, when the soil is subjected to a shallow depth of ponding at the surface. In hydrology, the term infiltration capacity is frequently used and is defined as the maximum rate at which rain can be absorbed by a soil in a given condition, i.e., it signifies soil infiltrability. It is this characteristic that determines how much of the failing rain will flow directly over the ground surface into streams and rivers and how much will enter the soil and will be retained there for some time before being either passed downward as percolation or returned to the atmosphere by the process of evapotranspiration. The terms infiltration capacity of the soil and infiltration rate are the same. Generally, under continues ponded conditions, the infiltration has a high rate in the beginning, decreasing rapidly and then more slowly until it approaches a constant rate.

As water moves deeper into the soil, filling the pores, the hydraulic gradient goes on decreasing and, hence the infiltration rate. Dispersion of aggregates, blocking of the cracks and channels, swelling of colloids also reduce the infiltration rate. Soil pores are also blocked and the infiltration rate is sharply reduced when water is turbid. The infiltration process is of great practical importance and it affects many aspects of hydrology and agriculture. It influences runoff, and determines the water content of the soil. It is possible to modify infiltration through suitable soil and crop management practices.

Profile Moisture Distribution During Infiltration

If we study the moisture distribution during infiltration in a uniform profile, it will be seen that the surface of the soil is saturated to a depth of a few centimeters. Below this saturated zone is a transmission zone through which water is being transmitted from the saturated zone to the wetting zone at the bottom. The transmission zone is a lengthening zone with uniform water content near saturation. In the wetting zone, the moisture content decreases with depth till it merges into the wetting front which forms a sharp boundary between the wet and dry soil. The wetting zone and wetting front move downward continuously during the process of infiltration. Thus, during infiltration, the wetted portion of the homogeneous soil profile comprises of five zones.

1. Transmission zone: In this zone, there is little change in water content from saturation. In general, the transmission zone with uniform water content. In this zone, the suction gradients are negligible and the water movement is primarily due to gravity.
2. Wetting front. This is the zone of the steep water content gradient and forms a sharp boundary between the wet and dry soil. Beyond this zone, there is no visible penetration of water.
3. Saturated zone: In this zone, the soil is saturated and extends to a depth of few millimeters (=1 cm).
4. Wetting zone: In this zone, the water content sharply decreases with depth.
5. Transition zone: In this zone, there is a rapid decrease of water content with depth to approximately few centimeters.

Consider the water distribution patterns during infiltration under shallow ponding of constant depth in three uniform soil profiles of sand, loam and clay. As can be observed, thereby requiring more water per unit volume to saturate it than the other two. Also, the wetting front is the sharpest in sand and the least so in clay, obviously, due to hydraulic characteristics at that suction value. Even though the depth of water standing on the surface (ponding) had been the same the sand profile was wetted throughout its entire 1.16 m depth within 4 hours, whereas loam and clay required 12 and 20 hours, respectively. Cumulative infiltration as a function of the square root of time shows that the curves

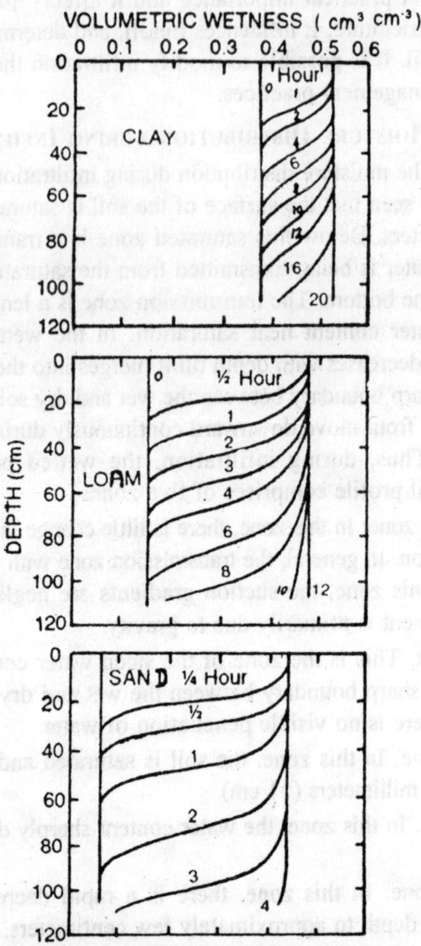

Fig. 3.8. Successive distribution of soil moisture during infiltration into uniform profiles of sand, loam and clay, all at an initial matric suction of 1/3 bar. Numbers on the curve indicate time.

are almost straight in the beginning as the pressure gradients predominate over the gravitational force.

However, as the profile gets wetted more deeply, the gravity force dominates. As this occurs, the infiltration rate approaches a constant value approximately equal to the hydraulic conductivity as saturation. Figure also shows that at a $t^{1/2}$ value of 150 hours, the sand, loam and clay profiles absorb nearly 65 cm, 25 cm and 7 cm water, respectively. In the profiles of sand, loam and clay mentioned above if we consider the

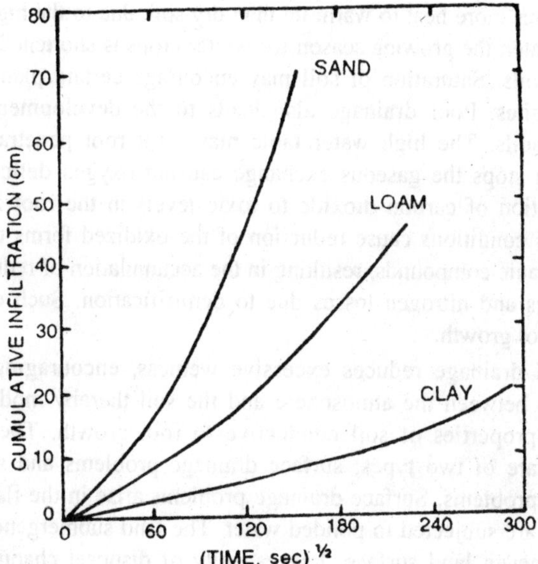

Fig. 3.9. Cumulative infiltration into uniform profiles of sand, loam, and clay (all at initial matric suction of 1/3 bar) as a function of the square root of time.

processes of infiltration and redistribution cumevaporation occurring simultaneously following a rainstorm, the water distribution patterns would be considerably different. It was seen that during a rainstorm, the upper horizons of sand remained unsaturated (rainfall rate < saturated hydraulic conductivity), while those of the finer textured soil profile approached saturation as the maximal rainfall rate nearly equaled or exceeded the hydraulic conductivity of these soils.

Total penetration of water during the first rainstorm duration (one-fourth of a day) was the deepest in the sand but did not exceed more than 0.5 m. During the 8-day redistribution-cum-evaporation phase the penetration of water was the shallowest in clay and the deepest in sand. Therefore, the upward moisture content gradients due to evaporation during the evaporation phase are steeper in clay than in coarse textured soil.

DRAINAGE

Drainage means removal of excess surface and subsurface water by means of some water-conveying devices. In agriculture we are more concerned with lowering of the water table and the relationship between the water table depth and the crop performance. Drainage is essential for crop production. The growth of crops is adversely affected by continuous soil saturation or by ponded water on the surface. Since wet soil requires more heat to warm up than dry soil, due to the high specific heat of water, the growing season for winter crops is shortened in poorly drained soils. Saturation of soil may encourage certain plant diseases and parasites. Poor drainage also leads to the development of salt-affected soils. The high water table may limit root penetration. Soil saturation stops the gaseous exchange causing oxygen deficiency and accumulation of carbon dioxide to toxic levels in the root zone. The anaerobic conditions cause reduction of the oxidized forms of organic and inorganic compounds, resulting in the accumulation of reduced toxic substances and nitrogen losses due to denitrification. Such conditions inhibit root growth.

Land drainage reduces excessive wetness, encouraging gaseous exchange between the atmosphere and the soil thereby modifying the physical properties of soil conductive to root growth. The drainage problem are of two types: surface drainage problems and subsurface drainage problems. Surface drainage problems arise in the flat areas of land that are subjected to ponded water. The land submergence may be due to uneven land surface, low capacity of disposal channels in the area and the outer conditions where the water surface may be above the ground level. The water available for ponding may come from rainfall, over irrigation, runoff, seepage or overflow from stream channels.

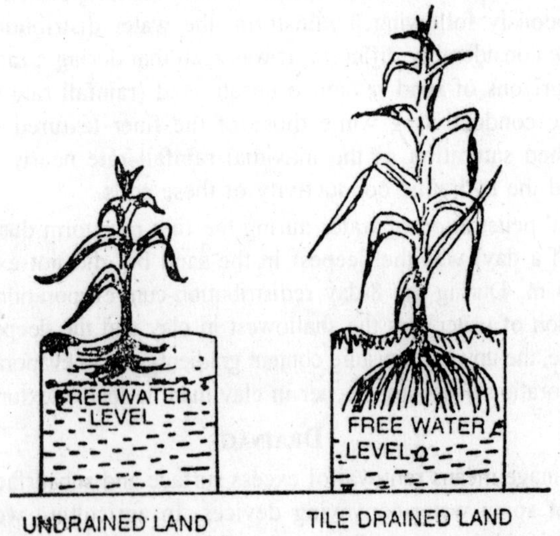

Fig. 3.10. Root and shoot development of crops grown on drained and undrained land.

Subsurface drainage problems may arise due to low soil permeability, humid climate and rise of ground water causing development of a shallow water table. The method used for land drainage can, therefore, be classified into two broad categories, viz. Surface drainage and surface drainage. In the case of surface drainage, water is removed directly from the land by land smoothing, land grading, banding or ditching. Land surfaces are also reshaped to climate ponding and to create slopes so as to induce gravitational flow over land and through channels to an outlet. The excess water can also be diverted from the land by diversion ditches, dykes, etc. In subsurface drainage, surface and subsurface drain are established in the soil profile to collect and convey excess ground water to gravity or pumped outlets. The discharge by permitting the pressure to drop induces the flow of excess ground water through the soil into the drains. The ranges of depth and spacing generally used for installing tile drains in the field are give an in Table 3.5.

Table 3.5 Average depth and spacing of the drains in various soil types.

Soil type	*Hydraulic conductivity cm/day*	*Spacing of drains, m*	*Depth of drains, m*
Clay	0.15	10-20	1-1.5
Clay loam	0.15-0.5	15-25	1-1.5
Loam	0.5-2.0	20-35	1-1.5
Fine sandy loam	2.0-6.5	30-40	1-1.5
Sandy loam	6.5-12.5	30.70	1-1.5
Peat	12.5-25	30-100	1-2

The drainage system can be designed using empirical criteria from data collected through evaluation of existing drainage systems or by theoretical analysis of the problem based on the physical laws of water movement and testing the theory through the evaluation of existing systems.

Evaporation from the Free Water Surface

Evaporation is the process by which water is changed form the liquid state to the vapour state. A water body is made up of H_2O molecules, each of which in the state of motion in different direction, and attracted to each other by a force that is inversely proportional to the product of their masses. Evaporation takes place when some molecules have attained enough kinetic energy to eject themselves from the water mass. Since the escaping molecules are attracted by other molecules underneath them, only those molecules which possess greater kinetic energy are able to escape from the water surface. The kinetic energy of the molecules escaping from the water surface decreases because of negative work done on it by the intermolecular force. Therefore, some water molecules continuously escape from the surface while others return and penetrate into the mass over those molecules returning.

At a high temperature, the water molecules attain greater Kinetic velocities and therefore, tend to fly of into the lower layers of the atmosphere resulting in greater evaporation. Analogous top heat, water and electricity flow, the process of evaporation is dependent on a driving force and a resistance. The motion of the water molecules escaping from the water surface produces a pressure which called vapour pressure. It is actually the partial pressure of the water vapour in the atmosphere, which a mixture of gases and water vapour. According to the ideal gas law, the partial gas pressure is directly proportional to the number of gas molecules present at constant temperature and volume of the mixture.

The difference in the vapour pressure between the layer from which water is evaporating and that of the turbulent atmosphere constitutes the driving force. The resistance is dependent on the nature of the vapour flow medium. It can be diffusion impedance to liquid and vapour flow in the system, the thickness of the laminar layer adjacent to the evaporating surface or the resistance caused by the turbulent air. Under the conditions of no turbulent air. Under the conditions of no turbulence or calm atmosphere, continuous evaporation causes an increase in the vapour pressure, eventually leading to the saturation of the atmospheric layers near the water surface and cessation of evaporation. Under such conditions condensation takes place and the saturation deficit is zero, i.e.,

$$(e^0 - e) = 0$$

where e^0 = saturation vapour pressure of water and e = actual vapour pressure of water.

Thus, evaporation and condensation are continuous processes. It is obvious that evaporation will be faster than condensation because the overlying air layers are often not saturated air is formed adjacent to the water surface, under isothermal conditions, the evaporation function analogous to Darcy's equation can be expressed as

$$E = -K'\frac{de}{dz}$$

where E = evaporation rate; z = thickness of the layer: and K′ = transfer coefficient.

It is seen that there exists a gradient in the vapour pressure in the air layer immediately above the film saturated with pressure vapour. The evaporation is a function of this vapour pressure gradient. Under absolutely calm atmospheric conditions, a situation may occur where the gradient is practically zero. Under normal conditions, the air movement causes mixing of air layers encouraging turbulence and convection, thereby reducing the vapour pressure and permitting further evaporation. The stronger the wind, the more effective is the turbulent action in the air, causing greater difference between the surface and overlying air temperatures, and the greater is the effect of convection. In 1802, Dalton considered all these factors and first enunciated the fundamental law of evaporation from the free water surface. This law states that evaporation will occur only if the actual vapour pressure of the air above the water surface is less than that at the water surface. Thus, the rate of evaporation is expressed as

$$E = (e_0 - e_d) f(U)$$

where E = rate of evaporation; e_o = mean vapour pressure at the water surface; e_d =mean vapour pressure in the air at some observational height above the water surface; and $f(U)$ = a function whose value depends on the horizontal wind velocity.

From Equation it is clear that if other factors remain constant evaporation is proportional to the wind speed and the vapour pressure deficit $(e_0 - e_d)$

Evaporation from the soil Surface

Evaporation from the soil surface is essentially the evaporation of water surrounding the soil particles as thin films and filling the pore spaces between them. Therefore, the atmospheric conditions that govern the evaporation from a free water surface will also govern the rate of evaporation from the soil surface. However, in the case of evaporation from the free water in bulk, the supply of water is not limited while the evaporation from the soil will frequently be affected due to insufficient supply of water in the soil. Furthermore, the water molecules escaping from the soil will have to overcome greater resistance due to the attraction of soil particles than while escaping from a free water surface.

When the water content of the surface soil is reduced below a threshold value, evaporation practically ceases. The foregoing discussion reveals that evaporation of water from the soil demands three basic physical requirements: First, a continuous supply of heart to change the state of water from liquid to vapour should be available. Secondly, a vapour pressure gradient between the soil surface and the surrounding atmosphere should be maintained. Thirdly, there must be a continuous supply of water from or through the soil profile at the surface.

The first two conditions, namely, supply of energy and removal of vapour are external to the evaporating body and are influenced by the meteorological factors such as air temperature, humidity, wind velocity and radiation, which together determine the evaporative demand of the atmosphere or evaporation potential. The evaporation potential is defined as the maximum evaporation rate from the surface of a free water in bulk under the given meteorological conditions. The third condition determines the maximum rate at which soil can transmit water to the plane of evaporation. The evaporation rate evaporation rate is thus determined either by the evaporation potential or by the soil's own ability to transmit water to the plane of evaporation under the given condition, whichever is lesser.

Physical Properties of Water

Water differs markedly from substances having closely related electronic structures, a point which we can illustrate by considering the series CH_4, NH_3, H_2O, HF and Ne. Each member contains 10 protons and 10 electrons, but the number of hydrogen atoms decreases along the series from methane to neon. For this sequence, the melting points are –184°, –78°, 0°, –92°, and –249°, respectively. A similar striking variation in boiling points occurs, viz., –161°, –33°, 100°, 19°, and –246°. The relatively high melting and boiling points for water, compared with substances having a similar electronic structure, are indicative of its strong intermolecular forces. In other words, thermal agitation does not easily disrupt the water-water bonding. This attraction between molecules is responsible for many characteristic features of water.

Hydrogen bonding

The strong intermolecular forces in water result from the structure of the H_2O molecule. The internuclear distances between the oxygen and each of the two hydrogens are approximately 0.99Å, and the H—O—H angle is about 105°. The oxygen atom is strongly electronegative and tends to draw electrons away from the hydrogen atoms. This leaves the oxygen with a partial negative charge (δ), while the two hydrogens become positively charged (δ^+). The positively charged hydrogens of a water molecule are electrostatically attracted to the negatively charged oxygen in two neighbouring water molecules. This leads to "hydrogen bonding" between water molecules, with an energy of about 4.8

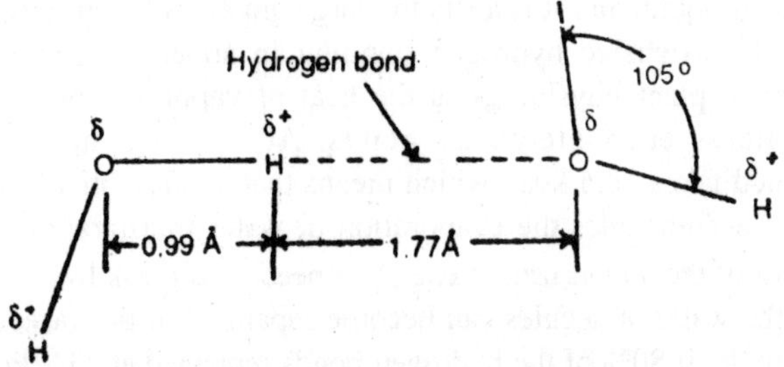

Fig. 3.11. Schematic structure of water molecules, indicating the hydrogen bonding resulting from the electrostatic attraction between the net positive charge on a hydrogen (δ^+) in one molecule and the net negative charge on an oxygen (δ^-) in a neighbouring water molecule.

kilocalories (kcal)/mole of hydrogen bonds. Such bonding of water molecules to each other leads to increased order in aqueous solutions.

In fact, liquid water becomes nearly crystalline in local regions, a condition which is extremely important for determining the molecular interactions and orientations that occur in aqueous solutions. Ice is a coordinated crystalline structure in which essentially all the water molecules are joined by hydrogen bonds. When heat is added so that the ice melts, some of these intermolecular hydrogen bonds are broken. The heat of fusion of ice at (0° is 80 gal/g or 1.44 kcal/mole. The total rupture of the intermolecular hydrogen bonds involving each of its hydrogens would require 9.6 kcal/mole of water. The heat of fusion thus indicates that (100) (1.44)/(9.6), or at most 15% of the hydrogen bonds are broken when ice melts (some energy is needed to overcome van der Waals attractions, so that less than 15% of the hydrogen bonds are actually broken upon melting).

Conversely, over 85% of the hydrogen bonds remain intact for liquid water at 0°. Since 1.00 cal is needed to heat 1 g of water 1°, (1.00 cal/g—°) (25°) (18 g/mole) ($^1/_{1000}$ kcal/cal), or 0.45 kcal/mole is required to heat water from 0° to 25°. If all of this energy were used to break hydrogen bonds, over 80% of the bonds would still remain intact at 25° Such bonding leads to the semicrystalline order found aqueous solutions. The extensive amount of intermolecular hydrogen bonds present in the liquid state contributes to the unique and biologically important properties of water that we will discuss throughout this chapter. The energy required to separate molecules from a liquid and move them into an adjacent vapour phase without a change of temperature is called the *heat of vapourization*. For water, the heat of vapourization at 100° is 539 cal/g or 9.71kcal/mole per gram, this is the highest heat of vapourization of any known liquid, and it reflects the large amount of energy required to disrupt the extensive hydrogen bonding in aqueous solutions. More pertinent to plant physiology is the heat of vapourization of water at temperatures encountered by plants. At 25° each mole of water evaporated takes 10.5 kcal, which means that a substantial heat loss by the plant accompanies the evaporation of water transpiration.

Most of the vapourization energy is needed to break hydrogen bonds so that the water molecules can become separated in the gaseous phase. For example, if 80% of the hydrogen bonds remained at 25°, then (0.80) (9.6), or 7.7 kcal/mole would be needed just to rupture them—other energy is needed to overcome the attractive van der Waals forces and for the expansion in going from a liquid to a gas. The heat loss upon evapouration of water is one of primary means of temperature regulation in land plants. It dissipates much of the energy gain resulting from the net radiation balance.

Surface tension

Water is further characterized by having an extremely high surface tension, which is evident at an interface between water and air. Surface tension can be defined as the force transmitted across a line in the surface, and its value at an air-water interface is 72.8 dynes/cm at 20°. Perhaps a better way of defining surface tension is to regard it as the amount of energy required to expand the surface by unit area. (Since a dyne is an erg/cm. note that surface tension has not only the dimensions of force per unit length but also the units of ergs/cm^2, or energy per unit area.) To see why energy is required to expand the water surface, let us consider water molecules which are brought from the interior of an aqueous phase to an air-water interface. If this involves only an expansion of the surface area—i.e., if there is no accompanying movement of other water molecules from the surface to the interior—then a loss in the water-water attraction intermolecular hydrogen bonds occurs with no compensating air-water attraction. Energy (72.8 ergs to increase the area by 1 cm^2) is thus needed to break the hydrogen bonds involved in moving water molecules from the interior of the solution to the air-water interface.

In fact the term "surface free energy" is more appropriate from a thermodynamic point view than is the conventional term "surface tension. The surface tension of an aqueous solution is ordinarily only slightly influenced by the composition of the adjacent gas phase, but it can be markedly affected by certain solutes. Molecules are far apart in gas—the density of dry air at 0° and 760 mm Hg is only 1.293 mg/cm^2 –

so that the frequency of collisions or other interactions between molecules in the gas phase and those in the liquid phase is relatively low. Certain solutes, such as sucrose of KCl, do not preferentially collect at the air-liquid interface and consequently have little effect on the surface tension of an aqueous solution.

On the other hand, fatty acids and certain lipids may become concentrated at interfaces and greatly reduce surface tensions. For example, 10 mM capric acid (a 6-carbon fatty acid) lowers surface tension about 15 dynes/cm below the value for pure water, while only 0.05 mM capric acid (a 10-C fatty acid) lowers it by about 25 dynes/cm. Substances such as soaps (salts of fatty acids) or denatured proteins with large hydrophobic side chains may collect nearly exclusively at the interface and reduce the surface tension of aqueous solutions to as low as 20 dynes/cm. A common feature of such "surfactant" molecules is that they have both polar and nonpolar regions.

Capillary rise

The intermolecular attraction occurring between like molecules in the liquid state, such as the water-water attraction based on hydrogen bonds, is referred to as *cohesion*. The attractive interaction between a liquid and a solid phase, such as between water and the walls of a small tube or capillary, is called *adhesion*. When the water-wall attraction is appreciable compared with the water-water cohesion, the walls are said to be *wettable*, and water then rises in the capillary. At the opposite extreme, where the intermolecular cohesive forces within the liquid are substantially greater than the adhesion between the liquid and the wall material, the upper level of the liquid in a capillary is lower than the surface of the free solution. Such capillary depression occurs for liquid mercury in glass capillaries.

For water in glass capillaries or in xylem vessels, the attraction between the water molecules and the walls is great, therefore the liquid rises. Since capillary rise has important implications in plant physiology, we will discuss its characteristics in a quantitative manner. As an example appropriate to the evaluation now water rise in a plant, let us consider a wettable capillary of inner radius *r* dipping into some aqueous solution. The strong adhesion of water molecules to the wettable wall leads to a rise of the fluid in the capillary. Since there is also a strong water-water cohesion in the bulk solution, water, is concomitantly pulled up into the lumen of the capillary as water rises along the walls. Viewed another way, the air-water surface greatly resists being stretched, a property reflected in the high surface tension of water at gas-liquid interfaces. This resistance tends to minimize the area of the air-water interface, a condition achieved if water also moves up in the lumen as well as along the sides of the capillary.

The tendency for water to rise along the walls of the capillary is thus transmitted to a volume of fluid. Let the height that the liquid rises in the capillary be *h* and the contact angle that the liquid makes with the capillary wall be α. The extent of the rise depends on α, so the properties of the contact angle will now be examined little more closely. The size of the contact angle depends on the magnitude of the liquid-solid adhesive force compared with that of the liquid-liquid cohesion. Specially, Young's equation indicates that the cohesion equals $(1 + \cos \alpha)/2$ times the cohesion. (Sometimes this relation is called the Young and Dupre equation—see Adamson, Bikerman, and/or Davies and Rideal for an authoritative treatment.) Consequently, when the adhesive force equals (or exceeds) the cohesive force in the liquid, $\cos \alpha$ must be unity, therefore the contact angle α equals zero ($\cos 0° = 1$). This is the case for water in capillaries made of clear smooth glass or having walls with polar groups on the exposed surface.

When the adhesive force equals half of the value of the cohesive force, $\cos \alpha$ is then zero and the contact angle is then 90° ($\cos 90° = 0$). In this case, the level of the fluid in the capillary is the same as that in the bulk of the solution. This latter condition is closely approached in water-polyethylene adhesion, where α equals 94°. As the liquid-solid adhesive force becomes relatively less compared with the intermolecular cohesion in the liquid phase, the contact angle increases toward 180° and capillary depression occurs, e.g., water has an α of about 110° for paraffin, while the contact angle for mercury interacting with a glass surface is near 150°. We can calculate the extent of the capillary rise by considering the balance of two forces : (1) gravity acting downward and (2) surface tension, which leads to a force in the upward direction in the case of wettable walls in a vertical tube.

The force pulling upward acts along the inside perimeter of the capillary, a distance of $2\pi r$, with a force per unit length equal to σ, the surface tension. However, the component of the force actually acting vertically upward is $2\pi r\sigma$, $\cos \alpha$ (where σ, is the contact angle). This upward force is balanced by the gravitational force acting on the liquid of density r and volume essentially $\pi r^2 h$. The gravitational force is simply the mass involved times the gravitational acceleration *g* (F = ma, Newton's second law of motion). In the present case gravity acts on a mass of fluid of approximately $\pi r^2 h\rho$, so the gravitational force is $\pi r^2 h\rho g$

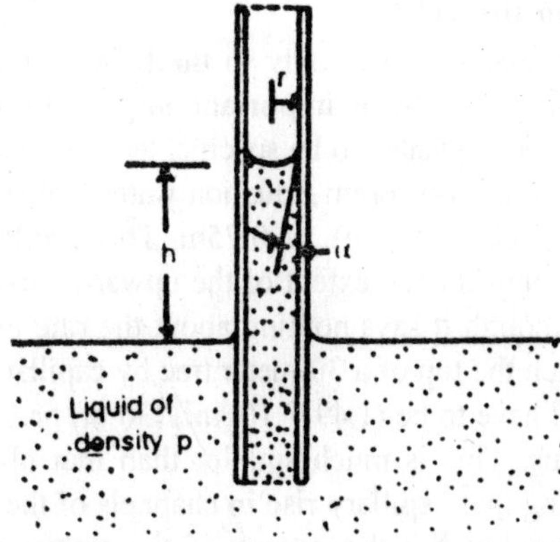

Fig. 3.12. Quantities involved in the capillary rise of a liquid.

acting downward. Since this force is balanced by 2πrσ, cos α pulling in the opposite direction, the extent of rise, h, is given by equating the two forces ($\pi r^2 h\rho g = 2\pi r\sigma, \cos\alpha$), which leads to

$$h = \frac{2\alpha \cos\alpha}{r\rho g}$$

Equation 23 indicates the readily demonstrated property that the extent of liquid rises in a capillary in inversely proportional to the radius of the tube. For water in glass capillaries as well as in many of the fine channels encountered in plants, where the cell walls have a large number of exposed polar groups, the contact angle can be near zero, in which case cos α appearing in Equation can be set equal to 1. The density of water is 0.998 g/cm^3 at 20° (actually, ρ in Equation is the difference between the liquid density and that of the displaced air, the latter being about 0.001 g/cm^3), and the acceleration due to gravity, *g*, is about 980 cm/sec^2. Using a surface tension for water at 20° of 72.8 dynes/cm, we obtain the following relation between the height of the rise and the capacity radius :

$$h_{(cm)} = \frac{0.149\, cm^2}{r_{(cm)}}$$

$$h_{(m)} = \frac{1.49 \times 10^{-5}\, m^2}{r_{(m)}}$$

where the subscripts (cm) or (m) in Equation mean that the dimensions involved are expressed in centimeters or meters, respectively.

Capillary rise in the xylem

Although Equations refer only to the height of capillary rise in a static sense, they still have important implications concerning the movement of water in plants. To be specific, let us consider a xylem vessel having a radius of 20 μm. From Equation water will rise in it to a height $(1.49 \times 10^{-5}\ m^2)/(20 \times 10^{-6}\ m)$, or 0.75m. This capillary rise would be sufficient to account for the extent of the upward movement of water in small plants, although it says nothing about the rate of such movement. For water to reach the top of a 30 meter tree by capillary action, however, the vessel would have to be $(1.49 \times 10^{-5} m^2)/(30\ m)$ or 5×10^{-7}m, which is 0.5 μm in radius. This is much smaller than that observed for xylem vessels, indicating that capillary rise in channels of the size of the xylem cells cannot account for the extent of the water rise in tall trees. Furthermore, the vascular system is not open to the air at the upper end, and thus xylem vessels are not really analogous to the capillary.

The numerous interstices in the cell wall of xylem vessels form a meshwork of many small tortuous capillaries, which can lead to an extensive capillary rise of water in a tree. A representative "radius" for these channels in the cell wall might be 5 nm (50 Å). Based on Equation a capillary of 5 nm radius could support a water column of $(1.49 \times 10^{-5})/(5 \times 10^{-9})$, or 3×10^3m—which is 3 kilometers (km)—far in excess of the needs of any plant. In other words, the cell wall could act as a very effective wick for water rise in its numerous small interstices, although the actual rate of such movement up a tree would generally be far from sufficient to replace the water lost by transpiration. Because of the appreciable water-wall attraction which can develop both at the top of a xylem vessel and in the numerous interstices of its cell walls, water that is already present in the lumen of a xylem vessel can be substained or supported at great heights.

The upward force, transmitted to the rest of the solution in the xylem vessel by water-water cohesion, overcomes the gravitational pull downward. Thus, the key feature for holding up water that is already present in the xylem vessel against the pull of gravity is the very potent attractive interaction (adhesion) of water to the cell wall surfaces at the top of the vessel. Two aspects of this warrant further consideration. First, what happens if the lumen of the xylem vessel becomes filled with air? Will water then refill it? The capillary rise of water is not sufficient to refill most xylem vessels greater than about 1 m in length, so that, in general, air-filled vessels are lost for conduction, such as the loss which occurs in the inner annual rings of most trees. The other aspect is more subtle, and we will consider it a little more extensively.

Tensile strength

The pulling on water columns which occurs in capillary rise and in the sustaining of water in the xylem requires that water be put under tension or negative pressure. In other words, water must have a definite tensile strength, the maximum tension (force per unit area) that it can withstand before breaking. In fact, the intermolecular hydrogen bonds resist the pulling apart of water in a column, if 80% of the hydrogen bonds are intact in water at 25°, their energy, will be (0.80) (9.6) or (7.7 kcal/mole)/(7.7kcal/mole)/(18 g/mole) or 0.43 kcal/g of water. For a density of 1.00 g/cm ^{2}, and replacing kcal by 4.18 × 10^{10} ergs, we calculate that the energy of the hydrogen bonds is 1.8 × 10^{10} ergs/cm^2 . The tension that is applied to a water column acts against this attractive energy of the hydrogen bonds. When the fracture is just about to occur at each hydrogen bond, the maximum possible tensile strength is developed. Thus, the maximum tensile strength would represent an input of energy of 1.8 × 10^{10} ergs/cm^3. Since an erg is a dynes-cm and a bar 10^6 dynes/ cm^2, the maximum tensile strength of water corresponds to 18,000 bars.

We might also attempt to calculate the maximum tensile strength of water by considering the electrostatic force of attraction between the δ^+ on the hydrogen of one water molecule and the δ^- on the oxygen of a neighbouring one. Dividing this force by the area of the water molecules involved, we could then calculate the tension developed—but such a approach involves formidable geometrical considerations. Experimentally, water withstands negative pressures (tensions) only up to approximately 300 bars 20° without breaking. (Observed tensile strength depend on the wall material, the diameter of the vessel in which the determinations are made, and any solutes present in the water sec Hayward.) Any local imperfections in the semicrystalline structure of water—such as those caused by H^+ and OH^- , which are always present, even in pure water—markedly reduce the observed tensile strength from the maximum value predicted based on hydrogen bond strengths.

Nevertheless, the measurable tensile strength for water provided by the intermolecular hydrogen bonds (up to 300 bars) is nearly 10% of that for copper or aluminium, and is sufficiently high to meet the demands encountered for water movement in plants. In summary, the great cohesive forces between water molecules allow an appreciable tension to exist in an uninterrupted water column in a wettable capillary or tube such as a xylem vessel. This property is important for the continuous movement of water from the root through the plant to the surrounding atmosphere during transpiration. In contrast to the stable situation for metals, water under tension is actually in a metastable state. If gas bubbles form in the water under tension in the xylem vessels, the column of liquid can be ruptured. Thus the introduction of another phase can disrupt the metastable state describing water under tension.

Minute air bubbles sometimes spontaneously from in the xylem as the external temperature rises. These usually adhere to the walls of the xylem vessel (or the pores of the perforation plate), and the gas in them slowly redissolves. However, if they grow large enough or a number of small bubbles coalesce, that xylem vessel could cease functioning. Moreover, freezing of the solution in xylem vessels can lead to bubbles in the ice (the solubility of air is quite low in ice), and these bubbles could then interrupt the water columns upon thawing. Some plants are in fact damaged by freezing and thawing of xylem sap and the consequent loss of water continuity in the xylem vessels.

Electrical properties

Another important physical characteristic of water is its extremely high dielectric constant, again a consequence of its molecular structure. A high dielectric constant for a solvent lowers the electrical forces between charged solutes. To quantify the magnitude of electrical effects occurring in a fluid, let us consider two ions having charges Q_1 and Q_2 and separated from each other by a distance r. The electrical force exerted by one ion on the other is expressed in Coulomb's law,

$$\text{electrical force} = \frac{Q_1 Q_2}{4\pi\varepsilon_o D r^2}$$

where ε_o is a proportionality constant known as the permittivity of a vacuum. In Equation D is a dimensionless quantity called the dielectric constant. It equals unity in a vacuum and is 1.00058 in air at 0° at a pressure of 760 mm Hg.

Any substance composed of highly polar molecules, such as water, generally has a high dielectric constant. Specifically, D for water is 80.2 and 20° and 78.4 at 25°, which are extremely high values for dielectric constants of liquids. By contrast, the dielectric constant of the non-polar liquid hexane is only 1.87—such a low value is typical of many organic solvents. Based on these two vastly different dielectric constants and Equation we can calculate that the attractive electrical force between ions such as Na^+ and Cl^- is (80.2/1.87), or 43 times greater in hexane than in water.

The much stronger attraction between Na^+ and Cl^- in hexane reduces the amount of NaCl that will dissociate, compared with the dissociation of this salt is aqueous solutions. Stated another way, the much weaker electrical forces in water, compared with those in organic solvents, permit a larger concentration of ions to be in solution. Water is thus a good solvent for charged particles. The electrostatic interaction between ions and water partially cancels or screens out the local electrical fields of the ions. Cations attract the negatively charged oxygens of the water molecule, while anions reciprocally attract its positively charged hydrogen atoms.

The oriented water molecules around the charged particles produce local electrical fields opposing the fields of the ions. The resultant screening diminishes the electrical interaction between the ions and allows more of them to remain in solution. This is the molecular basis for the high dielectric constant of water. The energy for the interaction between water and nonpolar molecules is usually less than the energy required to break water-water hydrogen bonds. Nonpolar compounds are therefore very soluble in water. Certain substances such as detergents, phospholipids, and proteins can have both a non-polar and also a polar region in the same molecule. In aqueous solutions these compounds can form aggregates, called *micelles*, in which the nonpolar groups of the molecules are in the center, while the charged or polar groups are external and interact with water. The lack of appreciable attractive electrostatic interaction between polar or charged species and the nonpolar (hydrophobic) regions of membrane lipids underlies the ability of biological membranes to limit the passage of such solutes into and out of cells and organelles.

Chemical Potential

The chemical potential of species j is a thermodynamic way of quantitatively describing the free energy associated with it and available for performing work. Since there are many concepts to be mastered before understanding such a statement, we will first briefly explore the concept of free energy.

Free energy and chemical potential

Plants and animals require a continual input of free energy. If we were to remove the sources of free energy, organisms would drift toward equilibrium and consequent cessation of life. The ultimate source of free energy is the sun. Photosynthesis converts its plentiful radiant energy, which is stored first in intermediate energy "currencies" like ATP and then in the altered chemical bonds which result when CO_2 and H_2O react to form a carbohydrate and O_2. When the overall photosynthetic reactions are reserved during respiration, the available free energy is reconverted into suitable energy currencies like ATP. In turn, such free energy is used to perform biological work such as transporting amino acids into cells, pumping blood powering the electrical machinery of the brain, or lifting weights. For every such process on both molecular and a global scale the free energy decreases.

In fact, the structure of cells, as well as that of ecosystems, is governed by the initial supply of free energy and by the inexorable laws of thermodynamics which describe the expenditure of that free energy. Instead of viewing the whole universe, the thermodynamicist focuses his attention on some small part of it referred to as a "system." Such a system can be potassium ions dissolved in water, a plant cell, a leaf, or even a whole plant. Something happens to the system; the potassium concentration is increased, some protein is made in a cell, a leaf abscises, or water moves up a tree. We then say that the system changes from state A to state B (the change may of course be brought about in many different ways). The minimum amount of work needed to cause the change from A to B can be defined as the free energy increase associated with it.

Alternatively, the free energy decrease in going from state B back to state A represents the maximum amount of work which can be derived from the transition. (If there were no frictional or other irreversible losses, we could do without the words "minimum" and "maximum.") Thus, the limits to the work done on or by a system when it undergoes a transition from one state to another concern change in free energy. Knowledge of the free energy under one condition compared with another allows us to predict the direction of spontaneous change or movement—a spontaneous change in a system at constant temperature and pressure always proceeds in the direction of decreasing free energy. Most systems of interest in biology are subject to a constant pressure (atmospheric) and remain at constant temperature, at least for short periods.

In discussing the energies of processes at constant temperature and pressure, the appropriate quantity is known as the *Gibbs free energy.* The Gibbs free energy has a very useful property in that it always decreases for a spontaneous process at constant temperature and pressure. Under such conditions, the decrease in Gibbs free energy equals the maximum amount of energy available for work, while if it increases for some transition, the change in Gibbs free energy represents the minimum amount of work required.

For our present purposes, we will simply note that the Gibbs free energy of an entire system is made up of additive contributions from each of the species present. We will therefore shift our emphasis to the individual components of the system and the manner in which their free energy can be described. It turns out that we can assign to every chemical component is a system a free energy per mole of the species. The quantity is called the chemical potential of species j and is given the symbol μ_j. We can view m*j* as a property of species j that tells how that species will react to a given change e.g., a transition of the system from state A to state B. During the transition the chemical potential of species j changes from μ_j^A a to μ_j^B. If μ_j^B is less than μ_j^A, then the free energy per mole of species j decreases. Such a process can take place spontaneously as far as species j is concerned—water flowing down hill is an example of such a spontaneous process. (We are restricting our attention to changes that occur when the system is at a constant temperature and subjected to a constant pressure.) In principle, we can harness the spontaneous change to do work. The maximum amount of work that can be done per mole of species j is $\mu_j^A — \mu_j^B$. Suppose now that μ_j^B equals μ_j^A. As far as species j is concerned, no work is involved in the transition from state A to state B—species j is then at equilibrium. (We will show later that not only are living systems as a whole far from equilibrium, but also many chemical species are not in equilibrium across cellular membranes). Finally, let us consider the case where μ_j^B is greater than μ_j^A. A change which increases chemical potential of a substance can only occur if some other change occurs in the system—this other change supplying the free energy required.

The pumping of blood along arteries using ATP to cause contraction of the heart muscles is an instance where the increase in chemical potential of a mole of water in the blood is accompanied by an even larger decrease in the free energy associated with ATP. The minimum amount of work that must be done per mole of species j to cause the energetically uphill transition is $\mu_j^B—\mu_j^A$. The conditions for a spontaneous change, namely a decrease in chemical potential, has important implications for discussing fluxes from one region to another. In particular, we can use the chemical potential difference between two locations as a measure of the "driving force" for the movement of that component. Thus, the bigger the chemical potential difference. $\mu_j^A—\mu_j^B$, the faster the spontaneous change takes place—in this case, the larger the flux of species j from region A to region B. The chemical potential of a substance clearly depends on its chemical composition, but it is also influenced by other factors.

Here we shall argue in an intuitive way about what environment factors might influence μ_j. To begin with, we note that chemical potential depends on the "randomness" (entropy) of the system, and concentration is a quantitative description of this randomness. For example, the passive process of diffusion describes neutral molecules spontaneously moving from some region to another where their concentration is lower. In more formal terms, diffusion of species j actually proceeds in the direction of decreasing μ_j, which in this case is the same as that of decreasing concentration. The importance of concentration gradients, or differences, in leading to the movement of a substance was also established in the discussion of diffusion. Here we will present the effect of concentration on chemical potential in a somewhat more sophisticated manner, using that part of concentration which is thermodynamically active.

Chemical potential also depends on pressure, which for situations of interest in pant physiology means hydrostatic pressure. The existence of pressure gradients can cause movements of fluids—the flow of crude oil in long-distance pipelines, blood in arteries, and sap in xylem vessels—which is another way of saying that pressure difference can affect the direction for spontaneous changes.

Electrical potential also affects the chemical potential of charged particles, hence it must be considered when predicting the direction of their movement. It is a matter of definition that work must be done to move a positively charged particle to a higher electrical potential. Accordingly, if an electrical potential difference is imposed across an electrolyte solution containing a uniform concentration of ions throughout, then cations will spontaneously move in one direction (towards the cathode, i.e., to regions of lower electrical potential), while anions will move in the opposite direction (toward the anode). It is clear, then, that in order to describe the chemical potential of a charged species we must include an electrical term in μ_j. Another contributor to chemical potential is gravity. We can readily appreciate that position in a gravitational field affects μ_j, inasmuch as work must be done to move the substance upward.

Although the gravitational term can be neglected for ion and water movements across plant cells and membranes, it is important for water movement in a tall tree. In sum, the chemical potential of some species should depend on its concentration, the pressure, the electrical potential, and gravity. We can compare the chemical potentials of a special on two sides of a barrier in order to decide whether or not the species is in equilibrium. If μ_j is the same on the two sides, we would not expect a net movement of species j to occur spontaneously across the barrier. We will use the relative values of the chemical potential of species j at various locations to predict the directions for spontaneous movement of that chemical substance (viz. l, toward lower μ_j), just as comparison of temperatures ;can be used to predict the direction for heat flow (toward lower T). Finally, we will find that $\Delta\mu_j$ from one region to another gives a convenient measure of the driving force on species j.

Analysis of chemical potential

For convenience and also because it has proved experimentally valid, we will represent the chemical potential of any species j by the following sum of the various components into which it can be analyzed:

$$\mu_j = \mu_j^* + RT\ 1n\ a_j + V_jP + z_jFE + mjgh$$

One measure of the elegance of a mathematical expression is the amount of usefulness or information which it contains. Based on such a criterion. Equation truly represents one of the most elegant relations in all of biology. After definition and describing the various contributions to μj indicated in Equation we will consider in greater detail the various terms for the important special case where species j is water.

Chemical potential, like electrical or gravitational potentials is a relative quantity. By this we mean that it must be expressed relative to some arbitrary level. Thus, an unknown additive constant, or reference level, μ_j^*, is included in Equation. Since it contains an unknown constant, the actual value of chemical potential is not determinable. For most applications of chemical potential in physiology, we are interested in the difference in the chemical potential between two particular locations, in which case only relative values of the chemical potential are important anyway. The arbitrariness of μ_j^* is of little consequence because it is added to each of the chemical potentials under comparison. Therefore μ_j^* cancels out when the chemical potential in one location is subtracted from that in another in order to obtain the chemical potential difference between the two locations.

Further illustrates the usefulness of considering differences in chemical potential instead of the actual value of μ_j. (We will return to μ_j^* when we consider μ_j in the standard state). Finally, we note that the units of μ_j and of μ_j^* are energy per mole of the substance, e.g., joules/mole or cal/mole. In certain cases we can adequately describe the chemical properties of species j by using the concentration of that solute, cj. Owing to molecular interactions, however, this usually requires that the total solute concentration below.

Molecules of solute species j interact with each other as well as with other solutes in the solution, and this influences the behaviour of species j. Such intermolecular interactions increase as the solution becomes more concentrated. The use of concentrations for describing the thermodynamic properties of some solute thus indicates an approximation, except in the limiting case of infinite dilution where interactions between solute molecules are negligible. Where exactness is required, activities which may be regarded as "corrected" concentrations—are employed. Consequently, for general thermodynamic considerations, such as Equation the influence of the amount of a particular species j on its chemical potential is not handled by its concentrations but by the activity of that species, a_j. The activity of solute j is related to its concentration by means of an activity coefficient, γ_j:

$$a_j = \gamma_j c_j$$

The activity coefficient is usually less than unity, since the thermodynamically effective concentration, or *activity* of a species is generally less than is its actual concentration. For an ideal solute, γ_j is one and the activity of species j then equals its concentration. This condition can be approached for real solutes in certain dilute aqueous solutions, especially for neutral species. Activity coefficients for charged species can be appreciably less than unity because of the importance of electrical interactions. The activity of a solvent is defined in a manner different from that for a solute. The aj of a solvent is defined as $\gamma_j N_j$, where N_j is the mole fraction of the solvent. An ideal solvent has $\gamma_{solvent}$ equal to 1, while an ideal solution has all activity coefficients equal to unity.

In the expression for chemical potential, Equation, the term involving the activity appears as +RT in a_j. Therefore the greater the activity of species j-loosely speaking, the higher its concentration—the larger will be its chemical potential. The appearance of the logarithmic form can be "justified" in a number of different ways, all of which are based on agreement with empirical observations. The factor RT multiplying in a_j in Equation gives the activity term the units of energy per mole. The term $V_j P$ in Equation represents the effect of pressure on chemical potential. Since essentially all measurements in plant physiology are made on systems subjected to atmospheric pressure, it is convenient to define P as the pressure in excess of this, and we will adopt such a convention here. V_j is the differential increase in volume of a system when a differential amount of species j is added, other species, temperature, pressure, electrical potential, and gravitational position remaining constant:

$$\overline{V}_j = \left(\frac{\partial V}{\partial nj}\right)_{niT, P, E, h}$$

The subscript n_i on the partial derivative in Equation means that the numbers of moles of all species present, other than the species j, are held constant when the derivative is taken—the four other subscripts are included to remind us of the additional parameters which must be held constant. V_j is called the *partial molal volume* of species j. (μ_j is actually the partial molal Gibbs free energy, $(\partial G/\partial n_j)_{ni}$ T.P, E, h. The partial molal volume of a substance is often nearly equal to the volume of a mole of that species, but because there is in general a slight change in total volume when substances are mixed V_j is not exactly equal to it. We will indicate that the value of V_j for water is about 18 cm²/mole when we specifically consider the chemical potential of water.

Since work is often expressed as pressure times change in volume, note that V_jP has the correct units for μ_j of energy per mole (V_j is the volume per mole). To help justify the form of the pressure term, let us imagine that the solution containing species j is built up by adding small volumes of species j while the system is maintained at a constant pressure P. The work done to add a mole of species j would then be the existing pressure times some volume change of the system which characterizes a mole of that species, viz. V_j. The influence of electrical potential on the μj of an ion is handled by the term z_jFE in Equation where z_j is an integer representing the charge number of species j, F is a constant known as the Faraday and E is electrical potential. Because water is uncharged ($z_w = 0$), the electrical term does not contribute to its chemical potential. However, electrical potential is of central importance when discussing ions and also the origin of membrane potentials. The full expressed for chemical potential, μ_j, given by Equation including z_jFE, is often referred to as *electrochemical potential* when discussing the properties of change particles.

For completeness, the gravitational term mjgh is included in Equation. It is apparent that to raise an object of mass mj per mole to a vertical height h requires an amount of work mjgh, where g is the gravitational acceleration (close to 980 cm/sec². In the case of water, which is the primary concern of this chapter, m_w is 18.016 g/mole. The gravitational term in the chemical potential is not only important for the fall of rain, snow or hail, but also it affects the percolation of water downward through porous soil and the upward movement of water in a tree. The gravitational term mjgh can have units of (g/mole)(cm²sec²) (cm) or (g-cm²)/sec²)(mole), which is ergs/mole.

Standard state

The additive constant term μ_j* in Equation is actually the chemical potential of species j for a suitable reference state. From the proceeding definitions of the various quantities involved, this reference state is attained when the activity of species j is unity (RT in a_j=0), the hydrostatic pressure equals atmospheric pressure (V_jP=0), the species is uncharged or the electrical potential is zero (z_jFE=0), we are at the zero level for the gravitational term (mjgh=0), and the temperature equals that for the system under consideration . Under these conditions μ_j equals μ_j*. An activity of unit is defined in different ways for the solute and the solvent.

To describe liquid properties, such as the dielectric constant, the heat of vaporization, and the boiling point, the most convenient standard state is that of the pure solvent. For a solvent, $a_{solvent}$ equals $\gamma_{solvent}$ times $N_{solvent}$ hence the activity is unity when the mole fraction $N_{solvent}$ is unity ($\gamma_{solvent}$ =1 for pure solvent). For example, the standard reference state for water is pure water at atmospheric pressure and at the temperature and gravitational level of the system under consideration. Water has an activity of unity when N_w as given by Equation is one—the concentration of water is then (1000 g of solvent)/(18.016 g/mole), or 55.5 molal. The accepted convention for a solute, on the other hand, is that *aj* is unity when $\gamma_j c_j$ equals 1 molal. For example, if γ_j equals one, a solution that has a 1 molal concentration of solute j has an activity of 1 molal for that solute. Thus, the standard state for an ideal solute is when its concentration is 1 molal, in which case RT 1n a_j is zero. The value of the chemical potential for a solute in the standard state, μ_j*, actually depends on the solvent.

In order to be explicit, let us consider a two-phase system consisting of water with an overlying layer of olive oil. It turns out that μ_j* for a polar molecule is smaller for an aqueous solution (where the species readily dissolves) than it is for an organic phase such as olive oil (where it is not as soluble). If we put the polar solute into our system, shake the water and olive oil together, and wait for the two phases to separate out again, the solute concentration will be higher in bottom or water phase. Let us next analyze this event using symbols, where phase A represents water and phase B is olive oil. Waiting for the phase separation is equivalent to waiting for the attainment of equilibrium, and so μ_j^A is then equal to μ_j^B.

In the present case this means that $\mu_j^{*,A}$ + RT In a_j^A equals $\mu_j^{*,B}$ + RT 1n a_j^B by Equation. But the polar solute is more soluble in water than it is in olive oil, hence RT In a_j^A is greater than RT 1n a_j^B. Because we at equilibrium, this means that μ_j^{*A} must be less than $\mu_j^{*,B}$, as we have already indicated. Such a two-phase separation procedure is actually followed in determining the olive oil-water partition coefficient, a_j^B/a_j^A (which is essentially the same as C_j^B/C_j^A for dilute solutions.) Moreover since the diffusion of many substances across membranes is correlated with their lipid solubility, such partition coefficients are of great utility in discussing the movement of solutes into and out of cells or organelles.

Hydrostatic pressure

Because of the rigid cell walls, fairly large hydrostatic pressures can exist in plant cells, while such pressures in animal cells generally are relatively small. In addition to being involved in the support of the plant, hydrostatic pressures are important for the movement of water and solutes in the xylem and probably also in the phloem. The term expressing the effect of pressure on the chemical potential of water is V_wP Equation, where V_w is the partial molal volume of water and P is the hydrostatic pressure in the aqueous solution in excess of the ambient atmospheric pressure. The approximate value of V_w can be estimated using the following argument. When 1 mole (18.0 g) of water is added to water, the volume increases by 18.0 cm 3, since the density of water at 20° is 0.99 g/cm^3. In other words, V_w is 18.0 cm^3/mole in the case of pure water. Although V_w (defined according to the partial derivative given in Equation can be influenced by the various solutes actually present, it generally is close to 18.0 cm^3/mole for a dilute solution, a value that we will use for calculations in this book. Various units are used for expressing pressures, atmospheres and bars being commonly employed in plant physiology.

A pressure of one atmosphere is defined as that pressure which can support either a column of mercury 76.0 cm. High or a column of water 10.3 m high. However, most current research concerning the water relations of plants uses pressures in bars. A bar is 10^6 dynes/cm^2, which equals 0.987 atmosphere. Pressure is force per unit area, which is dimensionally the same as energy per unit volume, e.g., one bar is 0.100 joule/cm^3 or 0.0239 cal/cm^3. Since V_w has the units of cm^3/mole, the quantity V_wP and hence m_w itself can be expressed in joules/mole or cal/mole.

Water activity and osmotic pressure

The presence of solutes in an aqueous solution tends to decrease the activity of water (a_w). As a first approximation, we can view the decrease in a_w as more and more solutes are added as simply a dilution effect. In other words, the concentration (or mole fraction) of water becomes less when the water molecules are displaced by those of solute. As its activity thus decreases, we would expect a lowering of the chemical potential of water, as does indeed occur (cf the RT In a_j term in μ_j). We also know that the presence of solutes can lead to an osmotic pressure (π) in the solution. An increase in the concentration of solutes raises the osmotic pressure, indicating that π and a_w change in opposite directions. In fact, the osmotic pressure and water activity are related, a fundamental definition of π being.

$$RT \ln a_w = -V_w \pi$$

where the subscript *w* refers to water. As absolutes are added, a_w decreases from its value of unity for pure water, ln a_w is therefore negative, and π (defined by Eq.) positive, which is consistent with the arguments presented above. Equational so indicates that RT In a_w, a term appearing in the chemical potential of water, can be replaced by $-V_w\pi$.

Unfortunately there is much variation on the use of the term "osmotic pressure" and "osmotic potential", as well as the algebraic sign of the quantity. Osmotic pressures were originally measured using an osmometer, a device employing a membrane that in the ideal situation was permeable to water but not to any of the solutes present. When pure water was placed on one side of the membrane and some solution on the other, there was a net diffusion of water toward the side with the solutes. To counteract this tendency and establish equilibrium, a hydrostatic pressure had to be applied to the solution side. This pressure was called the osmotic pressure—it depended of course on the presence of the solutes. Does an isolated solutions have an osmotic pressure? In the sense of requiring an applied hydrostatic pressure to maintain equilibrium, the answer is no.

Yet the same solution when placed in an osmometer can manifest an osmotic pressure. Thus, some physical chemistry texts have found it convenient to say that a solution has an osmotic *potential*—i.e., it could show an osmotic pressure if placed in an osmometer—and that osmotic potentials are a property characteristic of solutions. Many plant physiology books have come to a similar conclusion, e.g., Salisbury and Ross, but have defined the negative of the same quantity as the osmotic potential, in order to have the water potential (a quantity we will introduce shortly) depends directly on the osmotic potential and not on its negative. We will not try to resolve this dilemma. Instead, we will consistently refer to p defined by Equation as osmotic pressure. In principle, it is always possible to measure p as a pressure by using an osmometer. Using another "colligative" property is a direct and convenient way of evaluating osmotic pressure and water activity, as defined by Equation. The four colligative properties of a solution-those that depend on the concentration of solutes regardless of their nature, at least in a dilute enough solution—are the freezing point depression, the boiling point elevation, the lowering of the vapour pressure of the solvent, and osmotic pressure. Thus, if the freezing point of cell sap is measured, $\pi^{\text{cell sap}}$ can be calculated using the freezing point depression of 1.86° for a 1 molal solution and the *Van't Hoff relation.*

The Van't Hoff relation

For many purposes in plant physiology it is convenient to relate osmotic pressure directly to the concentration of solutes instead of expressing π in terms of the water activity, a_w, as is done in Equation. In general, the greater the concentration of solutes, the more negative In a_w becomes and the larger is the osmotic pressure. Thus, some way of expressing a_w in terms of the properties of the solutes would seem to be an appropriate approach for expressing π as a function of the various solute concentrations. The ensuring derivation not only will show how a_w can be expressed, but will also serve to indicate the many approximations necessary in order to achieve a rather simple expression for π. The activity of pure water is unity, while in general a_w equals $\gamma_w N_w$, where γ_w is the activity coefficient of water and N_w is its mole fraction. The mole fraction of a species is simply the ratio of the number of moles of that species to the total number of moles of all species in the system. Thus, N_w is given as:

$$N_W = \frac{n_W}{n_W + \sum_j ni} = \frac{\sum_j nj}{n_W + \sum_j nj}$$

where n_w is the number of moles of water, nj is the number of Moles of solute j, and the summation, $\sum_j$, is over all solutes in the system being considered. Equation also expresses the familiar relation that N_w equals 1 minus the mole fraction of solutes. For an ideal solution, γ_w equals 1. This condition of unit activity coefficient is approached for dilute solutions in which case n_w is much greater than $\sum_j nj$. (The expression $n_w > \sum_j nj$ can be taken as one way of defining a dilute solution, where the symbol > means "is much greater than".) Using Equation and assuming a dilute solution, we obtain the following relations for ln a_w:

$$\ln a_W \cong \ln N_W = \ln\left\{1 - \frac{\sum_j nj}{n_W + \sum_j nj}\right\} \cong -\frac{\sum_j nj}{n_W + \sum_j nj}$$

$$\cong -\frac{\sum_j ni}{n_w}$$

The penultimate step in Equation is based on the series expansion of a logarithm —ln (1 – x) = – x – $x^2/2$ –$x^3/3$– ..., a series which converges rapidly for | x | < 1 — and the last two steps employ the approximation

$$(n_w > \sum_j nj)$$

relevant to a dilute solution. Equation is thus restricted to dilute ideal solutions. Nevertheless, it is a useful expression which clearly indicates that in the absence of solute

$$(\sum_j nj = 0),$$

ln a_w is zero and a_w is unity while the presence of solutes decreases the activity of water from the value of one for pure water.

To obtain a rather familiar form for expressing the osmotic pressure, we can incorporate the approximation for ln a_ω given by Equation into the definition of osmotic pressure (RT ln a_ω = – $V_{\omega\pi}$, Equation which yields

$$\pi_S \cong -\frac{RT}{V\omega}\frac{\left(\sum_j nj\right)}{n\omega} = RT\sum_j \frac{nj}{V\omega\eta\omega} = RT\sum_j cj$$

where $V_{\omega\eta\omega}$ is the total volume of water in the systems (essentially the total volume of the system for a dilute aqueous solutions), nj/$V_{\omega\eta\omega}$ is the number of moles of species j in that volume and is therefore the concentration of species j(cj), and the summations in Equation are over all solutes. Osmotic pressure is often expressed by Equation, known as the Van't Hoff relation, but this justified only in the limit of dilute ideal solutions. An ideal solution has ideal solutes dissolved in an ideal solvent. Since in writing the string of equalities in Equation we assumed that γ_ω was essentially unity, our subsequently derived expression (Equation) strictly applies only when water acts as an ideal solvent (γ_ω =1.00).

To emphasize that we are neglecting any factors which cause Y_ω to deviate from unity and thereby affect the measured osmotic pressure (such as the interaction between water and colloids, which we will shortly discuss), π_s instead of π has been used in Equation, and we will follow this convention throughout the text. We must also consider the non-ideality of solutes. We can handle departures of the solutes from ideality by replacing c_j in Equation with aj, which is equal to $\gamma_j c_j$. Thus, the Van't Hoff relation for a dilute solution of real solutes can be represented by

$$\pi_s RT \sum_j \gamma_j c_j.$$

The effect of solute concentration on osmotic pressure as described by Equation was first clearly shown by the botanist W. Pfeffer in 1877. Both the measurement of osmotic pressures and recognition of their effects have been crucial for an understanding of the water relations of plants. Relatively dilute solutions have fairly large osmotic pressures.

The cellular fluid expressed from plants like pea or spinach often contain about 0.3 mole of osmotically active particles per kg or liter of water. This fluid is referred to as being 0.3 osmolal by analogy with molality, which refers to the total concentration, e.g., 0.1 molal $CaCl_2$ is nearly 0.3 osmolal since most of the $CaCl_2$ is dissociated. Note that molality (moles of solute/100 g solvent) is the correct unit for the Van't Hoff relation in its exact form, as well as for the other colligative properties. However, molarity (moles of solute/liter of solution, symbolized by M) is often a more convenient unit. Below 0.1M, molarity and molality are nearly the same for a small solute in an aqueous solution, while at high concentrations the molarity can be considerably less than the molality. At 20°, RT has the value of 24.37 liter-bars/mole. Using Equation, the osmotic pressure of an 0.3 osmolal solution is (24.37 liter-bars/mole) (0.3 mole/liter), or 7.3 bars (essentially 7 atmospheres). Thus, an 0.3 osmolal solution has a substantial osmotic pressure.

Matric potential

For certain applications in plant physiology, another term is frequently included in the chemical potential of water, viz. V_{wt}. Where τ is the metric potential, or matric pressure. The use of this term is sometimes convenient for dealing explicitly with interactions occurring at interfaces, although these interfacial forces can also be adequately represented by their contributions to π or P. In other words, the matric potential does not represent a new or different contribution to $\mu\omega$—it is only a book-keeping device for handling interfacial interactions. To help make this statement more meaningful, we will briefly consider the influence on the chemical potential of water of liquid-solid interfaces at the surfaces of colloids. When water molecules are associated with interfaces such as those provided by the surfaces of colloidal particles suspended in an aqueous solution, they have less of a tendency either to react chemically in the bulk solution or to escape into a surrounding vapour phase.

The presence of interfaces thus lowers the thermodynamic activity of the water (a_w), especially near the surfaces of the colloids. But the presence of solutes also lowers water activity an effect that can be expressed by Equation. As a useful first approximation, we can consider that those two different effects lowering water activity are additive in a solution which contains both ordinary solutes and colloids. Osmotic pressure (π) depends on the actual activity of water regardless of the reason for a_ω being different from unity, i.e., π still equals – (RT/V_ω) In a_ω, as it must by Equation. Combining these various ideas (and recalling that $a_\omega = \gamma_\omega N_\omega$) we can write the following relation :

$$\pi = -\frac{RT}{V_\omega}\ln \gamma_\omega N_\omega = -\frac{RT}{V_\omega}\ln\gamma_\omega - \frac{RT}{V_\omega}\ln N_\omega = \tau + \pi$$

where τ is a matric pressure resulting from the water-solid interactions at the surfaces of the colloids.

Note that π_s in Equation is to be identified with the osmotic pressure of all solutes, including the concentration of the colloids, as represented by Equation ($\pi_s \equiv RT\sum_j cj$). In other words, we were actually dealing only with—(RT/V_ω) ln N_ω in the derivation of π_s, since γ_ω was then set equal to unity. For an exact treatment when interfaces are –e.g., in the cytoplasm of a typical cell—we cannot set Y_ω equal to unity, since the activity of water, and hence the osmotic pressure (π), is strongly affected by the proteins and other colloids present. In such a case, Equation suggests a simple way in which matric pressure may be related to the reduction of the activity coefficient of water caused by the interactions at interfaces. (Note that if we let $\tau = -$ (RT/V_ω) ln γ_ω than γ_ω unity leads to a positive τ).

Equation should not be viewed as a relation during τ, rather, it is one example where it might be useful to represent interfacial interactions by a separate term which can be added to π_s, the effect of the solutes on π. Although π and a_ω may be the same throughout some system, both π_s and τ in Equation may vary. For example, water activity in the bulk of the solution may be predominantly lowered by the presence of the solutes, while at or near the surface of colloids the main factor decreasing a_ω from one could be the interfacial attraction and binding of water. This latter effect does not change the mole fraction of water but does reduce its activity coefficient, γ_ω. Other areas of plant physiology where matric potentials have been invoked are in descriptions of the chemical potential of water in soil or in cell walls.

The interfacial interactions between water and the many surfaces lining the interstices of the cell wall lead to a tension in the cell wall water. Such a tension is simply a negative hydrostatic pressure, i.e., P is less than zero. Although it is not necessary, P is sometimes used to refer only to positive pressures, while negative P in cell wall water is often defined as a positive matric potential. Some recent books define this same negative hydrostatic pressure in water in cell wall interstices or between soil particles as a negative matric potential. It is far more straightforward to refer to the quantity which reduces μ_ω in such pores as a negative P.

Water potential

From the definition of chemical potential (Equation) and the formal expression for osmotic pressure (Equation), we can write the chemical potential of water (μ_ω) as

$$\mu_\omega = \mu_\omega{}^* — V_{\omega\pi} + V_\omega P + mwgh$$

where the electrical term ($z_j FE$) is not included, since water carries no charge ($z_\omega = 0$). The quantity $\mu_\omega - \mu_\omega{}^*$ has proved to be of considerable importance for discussing the water relations of plants. It represents the work involved in moving one mole of water from some point in a system (at constant pressure and temperature) to a pool of pure water at atmospheric pressure, at the same temperature as the system under consideration, and at the zero level for the gravitational term. More to the point, a difference between two locations in the value of $\mu_\omega - \mu_\omega{}^*$ indicates that water is not in equilibrium, hence there will be a tendency for water to flow toward the region where $\mu_\omega - \mu_\omega{}^*$ is lower. A quantity proportional to $\mu_\omega - \mu_\omega{}^*$ and increasingly used in studies of plant water relations is the so-called water ψ. This is conveniently defined as

$$\psi = \frac{\mu_\omega - \mu_\omega}{V_\omega} = P - \pi + \rho_W gh$$

which follows directly from Equations plus the identification of m_ω / V_ω with the density of water, ρ_ω. Equations indicates that an increase in hydrostatic pressure raise the water potential, while an increase in osmotic pressure lowers it.

Some older terminology describing plant water relations can be identified from Equation. For example, —ψ is the quantity called "diffusion pressure deficit" or "suction pressure" (actually, the term $\rho_w gh$ was nearly at ways neglected). Although these expressions are quite colourful, the terms *water potential* (ψ) and *chemical potential of water* (μ_ω) are increasingly being used in biology and other disciplines, and we will use them in this text. Hydrostatic pressure (P) is often referred to as "turgor pressure" in considerations of the protoplast, or, from a complementary point of view, as "water pressure" in considerations of the pushing of the cell wall back against the protoplast. We have already mentioned that work must be performed to raise an object in the gravitational field of the earth. Thus, vertical position also affects the chemical potential, consequently, the term $\rho_w gh$ must be included in the water potential given by Equation at lease when water moves an appreciable distance vertically in the gravitational field.

In plant physiology the magnitude of $\rho_w g$ is conveniently expressed as 0.098bar/m—if water moves 10 m vertically upward in a tree, the gravitational contribution to the water potential is increased by about 1 bar. For most applications in this text, such as consideration of chemical reactions or the crossing of membranes, there is no change in vertical position and the gravitational term will generally be omitted from μ_j and ψ.

Central Vacuole and Chloroplasts

The water relations of both the large central vacuole and the chloroplasts will be considered next, using the water potential as defined above. We will focus our attention on the situation where there is no net flow of water across the limiting membranes surrounding these subcellular compartments and for the special case of non-penetrating solutes. The central vacuole occupies up to 90% of the volume of a mature plant cell, which means that under such circumstances most of the cellular water is in the vacuole.

The vacuolar volume is generally from 10^4 to 10^5 μm^3 for the mesophyll cells in a leaf, but can be much larger in certain algal cells. Because its size is similar to that of the cell, the vacuole cannot be easily removed from the plant cell without rupturing its surrounding membrane, the tonoplast. In other words, procedures which remove the cell wall and open up the plasmalemma usually also break the tonoplast, especially for small cells. Chloroplasts are much smaller than the central vacuole, often having volumes near 30 μm^3 in *vivo* (although the sizes vary considerably with the plant species). When chloroplasts are carefully isolated, suitable precautions will ensure that their limiting membranes remain intact. Such intact chloroplasts can be placed in solutions having various osmotic pressures and the resulting movement of water into or out of the isolated plastids can be precisely measured.

Water relations of the central vacuole

In order to predict whether and in what direction water will move, we need to know the value of the water potential in the various compartments under consideration. At equilibrium the water potential is the same in all communicating phases, such as those separated by membranes. For example, when waters is in equilibrium across the tonoplast, the value of the water potential is the same in the vacuole as it is in the cytoplasm. No force then drives water across this membrane, and no net flow of water occurs into or out of the vacuole. The tonoplast most likely does not have an appreciable difference in hydrostatic pressure across it.

A higher internal hydrostatic pressure would cause a typical slack (folded) tonoplast to be mechanically pushed outward. The observed lack of motion thus indicates that ΔP is zero across a slack tonoplast. If the tonoplast were taut, ΔP would cause a stress in the membrane, analogous to the cell wall stresses discussed in the last chapter. For a tonoplast 10 nm thick surrounding a spherical vacuole 10 μm in radius, the stress would be $r\Delta P/2t$, which is $(10 \times 10^{-6}\text{m})\ \Delta P/[(2)(10\text{–}10^{-9}\ \text{m})]$, or $500\Delta P$. If the hydrostatic pressure were only 0.01 bar higher in the vacuole compared with the cytoplasm, in a taut tonoplast there would then be a stress of (500)(0.01), or 5 bars. This exceeds the tensile strength of biological membranes — most membranes rupture before a stress of even 1 bar develops in them. Thus, P may well have essentially the same value in the cytoplasm and the vacuole.

With this simplifying assumption, and for the equilibrium situation where ψ is the same in the two phases, Equation ($\psi = P - \pi + \rho_w gh$ where $\Delta h = 0$ across a membrane) gives $\pi^{cytoplasm} = \pi^{vacuole}$, where $\pi^{cytoplasm}$ is the osmotic pressure in the cytoplasm and $\pi^{vacuole}$ that in the central vacuole. The vacuole appears to be a relatively homogeneous aqueous phase, while the cytoplasm is a more complex phase containing many colloids and membrane-bounded organelles. Since the vacuole contains few colloidal or other interfaces, any matric potential in it is probably negligible compared with the osmotic pressure resulting from the vacuolar solutes. Expressing osmotic pressure by Equation ($\pi = \pi_s + \tau$) and assuming that $\tau^{vacuole}$ is negligible, we can replace $\pi^{vacuole}$ in Equation by $\pi_s^{vacuole}$, i.e, the decrease in vacuolar water activity is due almost solely to the presence of solutes. On the other hand, water activity in the cytoplasm is presumably considerably lowered both by solutes and by the presence of interfaces. In other words, individual water molecules in the cytoplasm are never very far away from proteins, organelles, or order colloids, so that interfacial interactions can appreciably affect $a_w^{cytoplasm}$.

In terms of Equation this means that both $\pi_s^{cytoplasm}$ and $\tau^{cytoplasm}$ contributes to $\pi^{cytoplasm}$. Hence, Equation indicates that $\pi_s^{cytoplasm} + \tau^{cytoplasm}$

equals $\pi_s^{cytoplasm}$. Since $\tau_s^{cytoplasm}$ is a positive quantity, we conclude that at equilibrium, the osmotic pressure in the vacuole due to solutes, $\pi_s^{vacuole}$, must be larger than $\pi_s^{cytoplasm}$. Equation leads then to the prediction that the vacuole has a higher concentration of osmotically active solutes than does the cytoplasm. There are various speculations concerning the role of the large central vacuole in plant cells. The vacuole might act as a storage site for metabolites, ions, or toxic products. Because of the large central vacuole, the cytoplasm occupies a thin layer around the periphery of the cell. Therefore, for its volume, the cytoplasm has a relatively large surface area (across which diffusion can occur). Vacuoles also provide large, relatively simple compartments in which hydrostatic pressures lead to the cellular turgidity necessary for support of the plant. The actual function of the central vacuole most likely entails a combination of these various possibilities.

Boyle-Van't Hoff relation

The volume of a chloroplast or other membrane-bounded body changes in response variations in the osmotic pressure of the external solution, π°. This is a consequence of the properties of membranes, which generally allow water to move readily across them, at the same time restricting the passage of certain solutes, such sucrose. For example, if π° outside a membrane-bounded aqueous compartment were raised, we would expect that water would flow out across the membrane to the region of lower chemical potential and thereby decrease the compartment volume, while in general there would be very little movement of solutes in such a case. This differential permeability leads to the "osmometric behaviour" characteristic of many cells and organelles. The conventional expression quantifying this volume response to changes in the external osmotic pressure is the Boyle-Van't Hoff relation:

$$(\pi^\circ V - b) = RT\sum_j \phi_j n_j = RTn$$

where π° is the osmotic pressure of the external solution, V is the volume of the cell or organelle, φ is the so-called non-osmotic volume (osmotically inactive volume) and is frequently considered to be the volume of a solid phase within volume V which is not penetrated by water, n_j is the number of moles of species *j* within V–b, the osmotic coefficient (φ_j) is generally construed as a correction factor, and

$$n = \sum_j \phi_j n_j$$

is the apparent number of osmotically active moles in V–b. In this chapter we will derive the Boyle-Van't Hoff relation using the concept of chemical potential, and we will extend the treatment by employing irreversible thermodynamics.

Although the derived Boyle-Van't Hoff expression will be used only to interpret the osmotic responses of chloroplasts, the equations that will be developed are quite general and can be applied equally well to data obtained with mitochondria, whole cells, or other membrane-surrounded bodies. The Boyle-Van't Hoff relation applies to the equilibrium situation when equality of the water potential has been achieved across the membranes surrounding chloroplasts exposed to a particular external osmotic pressure. When ψ^i equals ψ^o net water movement across the membrane ceases, and the volume of a chloroplasts has a steady value. (The superscript *i* refers to the inside of the cell or organelle, the superscript *0* designates the outside.) If we were to measure the chloroplast volume under such conditions, the external solution would generally be at atmospheric pressure ($P^o = 0$ bars). From Equation ($\psi = P - \pi$), the water potential in the external solution is then

$$\psi^o = -\pi^o$$

As is the case for ψ^o, the water potential inside the chloroplasts depends on osmotic pressure. In addition, the internal hydrostatic pressure (P^i) may be different from atmospheric pressure and should be included in the expression for ψ^i. Macromolecules and solid-liquid interfaces inside the chloroplasts can lead to a matric potential, τ^i. To allow for this possibility, the internal osmotic pressure (π^i) will be considered as the sum of the solute and the interfacial contributions, in the manner expressed by Equation. Using Equation ψ^i is therefore

$$\psi^i = P^i - \pi^i = P^i - p_s{}^i - \tau^i$$

We will now re-express π^i_s and then equate ψ^o to ψ^i. For a dilute ideal solution inside the chloroplasts, $\pi_s{}^i$ equals

$$RT \sum_j n^i_j / (V_w n^i_w)$$

by Equation. The solutions of real solutes necessitate the introduction of an activity coefficient for solute j, γ_j, which approaches one as the solution becomes more dilute. To handle deviations from ideal behaviour for the solutes, n_j^i can thus be replaced by $\gamma_j^i n_j^i$. Make these changes and equating the water potential outside the chloroplast to that inside the condition for water equilibrium across the limiting membranes of the chloroplast becomes

$$\pi^\circ = RT\frac{\sum_j \gamma_j^i n_j^i}{V_w n_w^i} + \tau^i - P^i$$

Although this expression may look a little frightening, it states only that at equilibrium the water potential is the same on the two sides of the membranes and that we can explicitly recognize various contributors to the two ψ_s involved.

To appreciate the refinements this thermodynamic treatment introduces into the customary expression describing the osmotic responses of cells and organelles. Equation should be compared with the conventional Boyle-Van't Hoff relation. The volume of water inside the chloroplast is $V_w n_w^i$, since n_w^i is the number of moles of internal water and V_w is the volume of one mole of water. This factor in Equation can be identified with V–b in Equation.

Instead of being designated the "non osmotic volume," b is more appropriate v called the "non water volume," for it includes the volume of the internal solutes including that of colloids. In other words, the total volume (V) minus the non water volume (b) equals the volume of internal water ($V_w n_w^i$). The osmotic coefficient, ϕ_j in Equation is generally less than *l*, as is the activity coefficient of an alternative solute, γ_j^i, in Equation. But γ_j^i cannot simply be equated to ϕ_j, in Equation. But *Yj* cannot simply be equated to fj since other factors are also involved. Finally the possible hydrostatic and matric contributions included in Equation are neglected in the usual Boyle-Van't Hoff relation. In short, although certain approximations and assumptions have crept into Equation (e.g., that solutes do not cross the limiting membranes, and using dilute solution considerations), it is still far more satisfactory expression for describing osmotic responses than is the conventional Boyle-Van't Hoff relation.

Osmotic responses of chloroplasts

To illustrate the use of Equation in interpreting osmotic data, we will consider osmotic responses of pea chloroplasts suspended in external solutions of various osmotic pressures. Since it is customary to plot the volume V versus the reciprocal of the external osmotic pressure, $1/\pi^\circ$, certain algebraic manipulations are needed in order to put Equation into a more convenient form for discussing such osmometric studies. By transferring $\tau^i - p^i$ to the left side of Equation and then multiplying both sides by $V_w n_w^i/(\pi^\circ - \tau^i + P^i)V_w n_w^i$ can be shown to equal

$$RT\sum_j \gamma_j^i n_j^i / (\pi^\circ - \tau^i + P^i).$$

The measured chloroplast volume (V) can be represented by $V_w n_w^i$ + b, i.e., as the sum of the aqueous and non aqueous contributions. Hence, V should be equal to

$$RT\sum_j \gamma_j^i n_j^i / (\pi^\circ - \tau^i = P^i) + b,$$

if the modified form of the Boyle-Van't Hoff relation is obeyed. Figure indicates that the volume of pea chloroplasts varies linearly with $1/\pi^\circ$, over a considerable range of external osmotic pressures. Consequently, $\tau^i = P^i$ occurring in both Equation and the above expression must either be negligibly small for pea chloroplasts, compared with the various values of π° employed, or perhaps proportional to π°. Actually, the situation is rather complicated, since both τ^i and γ_j^i depend on the chloroplast volume. For simplicity, we will ignore these effects and consider only the observed proportionality between V and l/π°. In other words, we will assume that V–b (or, $V_w n_w^i$) equals

$$RT\sum_j \gamma_j^i n_j^i / \pi^\circ$$

for pea chloroplasts. We will return to such considerations, where we will further refine the Boyle-Van't Hoff relation to include the more realistic case in which solutes can cross the limiting membranes.

The relatively simple measurement of the volumes of pea chloroplasts for various external osmotic pressures can yield a considerable amount of information about the organelles. If we measure the volume of the isolated chloroplasts at the same osmotic pressure they encounter in the cytoplasm, we could then determine the volume the organelles have when in the plant cell. Cell sap expressed from pea leaves was, found to have an osmotic pressure of 7.0 bars. Such "sap" comes mainly from the central vacuole, but since we expect $\pi^{cytoplasm}$ to be essentially equal to $\pi^{vacuole}$ Equation, $\pi^{cell\ sap}$ is about the same as $\pi^{cytoplasm}$ (a shall uncertainty exists inasmuch as during extraction the cell sap can come into contact with water in the cell walls).

At an external osmotic pressure of 7.0 bars, pea chloroplasts have a volume of 29 μm^3 when isolated from illuminated plants, and 35 μm^3 when isolated from plants in the dark. Since these volume occur at approximately the same osmotic pressure found in the cell, they are presumably reliable estimates of pea chloroplast volumes in *vivo*. The data clearly indicate that light affects the chloroplasts in many plants do have a larger volume in the dark than in the light. The decrease in volume upon illumination of the plants is observed by both phase contrast and electron microscopy as a flattening or decrease in thickness of the chloroplasts. This flattening, which amounts to about 20% of the thickness for pea chloroplasts in vivo, is in the vertical direction for the chloroplast. We also note that the slopes of the osmotic response curves are equal to

$$RT\sum_j \gamma_j^i n_j^i.$$

The slope is 50% greater for pea chloroplasts from plants in the dark than from those in the light, suggesting that chloroplasts in the dark contain more osmotically active particles. In fact, illumination causes a marked efflux of K^+ and Cl^- from the chloroplasts. Also, some Mg^{++} enters the chloroplasts in the light, perhaps by an exchange of one Mg^{++} for two K^+. Such a process would also reduce

$$\sum_j \gamma_j^i n_j^i,$$

since only one Mg^{++} would enter for every two K^+ that leave. The entry of Mg^{++} apparently causes its concentration to rise by about 10 mM in the chloroplasts, which may act as a control mechanism for photophosphorylation and CO_2 fixation, both of which depend on magnesium. Although such speculation takes us far away from our present consideration of water, it indicates the kind of insight that can be obtained from osmometric studies.

The intercept on the ordinate is the chloroplast volume theoretically attained in an external solution of infinite osmotic pressure—$1/\pi°$ equaling zero is the same as a $\pi°$ of infinity. For such an infinite $\pi°$, all of the internal water would be removed ($n_w{}^i = 0$ moles), and the volume, which is actually obtained by extrapolation, would be that of the non-aqueous components of the chloroplasts. Thus, the on the ordinate of a V versus $1/\pi°$ plot corresponds to *b* in the conventional Boyle-Van't Hoff relation (Equation). This intercept equals 17 μm^3 for chloroplasts both in the light and dark. (The extra solutes in a chloroplast in the dark correspond to less than 0.1 μm^3 of solids). Using the chloroplast volumes obtained for a $\pi°$ of 7.0 bars and 17 μm^3 for the non-water volume *b*, we find that the water content of pea chloroplasts in the dark is [(35 μm^3–17 μm^3)/(35 μm^3)](100) or 51% by volume, and in the light? (29-17)/29?(100) or 41%. These relatively low water contents in the organelles are consistent with the extensive internal lamellar system and the abundance of CO_2 fixation, enzymes in chloroplasts. Thus, osmometric responses of certain cells and organelles can be used to provide information on their fractional water content.

Water Potential and Plant Cells

In this section we will shift our emphasis from a consideration of the water relations of subcellular bodies to those of whole cells, and extend the development to include the case of water fluxes. Whether water enters or leaves a plant cell, and how much depends on the water potential outside compared with that inside. The external water potential (ψ^o) can often be varied experimentally, and the direction as well as the magnitude of the resulting water movement will give information about ψ^o. Moreover, the equilibrium value of ψ^o can be used to estimate the internal osmotic pressure, π^i. A loss of water from plants—indeed, sometimes also an uptake— occurs at cell-air interfaces. As we would expect, the chemical potential of water in cells compared with that in the adjacent air determines the direction for net water movement at such locations. Thus, we must obtain an expression for the water potential in a vapour phase, and then relate this ψw^v to ψ for the liquid phases in a cell. We will specifically consider the factors influencing the water potential in the component at plant cell-air interface, viz., the cell wall. We will find that $\psi^{cell\ wall}$ is dominated by a negative hydrostatic pressure resulting from surface tension effects in the cell wall pores.

Incipient plasmolysis

For usual physiological conditions, a positive hydrostatic pressure exists inside a plant cell, i.e., it is under turgor. By suitably adjusting the solutes in an external solution that barthes such a turgid cell, it is possible to reduce P^i to zero and thereby to obtain an estimate of π^i, as the following argument will indicate. Let us place the cell in pure water (π^o=0 bars) at atmospheric pressure (P^o = 0 bars). ψ^o is than zero (ignoring the gravitation term, $\psi = P - \pi$, Equation while ψ^i is $P^i - \pi^i$ (the inner phase can be the cytoplasm). If the cell can be assumed to be in equilibrium with this external solution, $\psi^o = \psi^i$, P^i must then equal π^i. Suppose that π^o is now gradually raised from its equal initial zero value. The osmotic pressure outside the cell can be increased for example by adding solute to the external solution. If the cell remains in equilibrium, then π^o equals $\pi^i - P^i$ (P^o= 0 bars, since the external solution is at atmospheric pressure). As π^o is increased, P^i will tend to decrease, while usually does not change very much.

More precisely, because the cell wall is quite rigid, a plant cell will not change its volume appreciably in response to changes in P^i. (Since the cell wall has elastic properties some water flows out as P^i decreases and the cell shrinks, a point which we will return to shortly.) Therefore, if the cell volume can be assumed not to change in response to changes in P^i and no internal solutes leak out, π^i will remain constant. As the external osmotic pressure is raised, π^o will eventually become equal to π^i. In such a plant cell at equilibrium, P^i is zero, which means that no internal hydrostatic pressure is exerted against the cell wall when π^o equals πi. The cell would thus lose its turgidity. If π^o were increased even further, an appreciable amount of water would flow out of the cell and plasmolysis would occur as the plasmalemma pulls away from the rigid cell wall. Ignoring for the moment any overall volume changes of the cell, we find that the condition under which water just begins to move out of the cell—referred to as the point of incipient plasmolysis,

$$\pi^o_{\text{plasmolysis}} = \pi^i$$

Equation suggests the only a relatively simple measurement ($\pi^o_{\text{plasmolysis}}$) is needed in order to estimate the osmotic pressure (π^i) occurring inside an individual plant cell.

The existence of an internal hydrostatic pressure within a plant cell leads to stresses in its cell wall and a resulting elastic deformation or strain. The decrease of P^i to 0 bars—which occurs in taking a turgid plant cell to the point of incipient plasmolysis—must therefore be accompanied by a contraction of the cell as the wall stresses are removed. This decrease in volume means that some water will actually flow out of the cell before the point of incipient plasmolysis is reached. If no internal solute leaves as the cell shrinks, then the osmotic pressure inside will increase (the same amount of solute in a smaller volume).

As a useful first approximation, we can assume that the increase in osmotic pressure reciprocally follows the change in volume, i.e., the product of π^i and the cellular volume is approximately constant. With this assumption, the osmotic pressures in cells—described by Equation, and determined by using the technique of measuring the point of incipient plasmolysis—can be corrected to its original value by using the ratio of the initial volume to the final volume of the cell. The change in volume is only a few percentage points for some plant cells, in which case fairly accurate estimates of p^i can be obtained from plasmolytic studies alone. Measurement of $p^o_{\text{plasmolysis}}$ often give values near 7 bars for cells in storage tissues like onion bulbs or carrot roots and in leaves like those of pea or spinach. These values of the external osmotic pressure provide information on various contributors to the water potential inside the cell, as Equation indicates. For the case in which water is in equilibrium within the plant cell at the point of incipient plasmolysis, the term π^i in Equation is the osmotic pressure both in the cytoplasm and in the vacuole.

Under certain circumstances it is convenient to replace π^i by $\pi^i_s + \tau^i$ (Equation), where π_s^1 is the osmotic pressure contributed by the internal solutes and τ^i is the matric potential. This relation was invoked for example when the implications of Equation ($\pi^{cytoplasm} = \pi^{vacuole}$) were discussed, and the various arguments presented at that time extend to the present case. Specifically, since the possible matric potential in the vacuole is probably negligible, $\pi^o_{plasmolysis}$ should be a rather good estimate of $\pi_s^{vascuole}$. Moreover $\pi^{cytoplasm}$ is most probably significant, as the water activity in the cytoplasm can be lowered by the many interfaces present. Thus, $\pi^o_{plasmolysis}$ is an upper limit for $\pi_s^{cytoplasm}$, We see, therefore, that determination of the external osmotic pressure at the point of incipient plasmolysis provides information on the osmotic pressure existing in different compartments within the plant cell.

Activity and water potential of water vapour

Water molecules in an aqueous solution continually escape into a surrounding gas phase. At the same time, water molecules also condense back into the liquid. The rates of escape and condensation depend inter *alia* on the chemical activities of water in the gas and the liquid phases the two flows becoming equal at equilibrium. The gas phase adjacent to the given solution then contains γ, much water as it can hold at that temperature and still remain in equilibrium with the liquid. It is thus saturated with water vapor for the particular solution under consideration.

The partial pressure exerted by the water vapor in equilibrium with pure water is known as the *saturation vapour pressure*. The vapour pressure at equilibrium depends on the temperature and the solution but is independent of the relative or absolute amounts of liquid and vapor. When air adjacent to pure water is saturated with water vapour (100% relative humidity), the gas phase then has the maximum water vapour pressure possible at that temperature—unless it is supersaturated, a non-equilibrium situation. This saturation vapour pressure of pure water increases markedly with temperature, e.g., it is equivalent to 4.6 mm of Hg at 0° and 55.3 mm at 40°. Thus, air of 100% relative humidity at 0° would be only (100)(4.6)/(55.3), or 8% relative humidity when heated at constant pressure to 40°. As solutes are added to the liquid phase and the activity of its water is thereby lowered. water molecules have less of the tendency to leave the solution, so that the water vapor pressure in the gas phase at equilibrium becomes less—this is one of the colligative properties of solution that we mentioned earlier.

In fact, for dilute solution the actual partial pressure of water vapour (P_{wv}) at equilibrium depends linearly on the mole fraction of water (N_ω)

in the liquid phase—this is *Raoult's law*. For pure water, Nw equals one and Pwv has its maximum value, viz., P_{wv}*, the saturation vapour presents. The partial pressure of the water vapour determines the activity of water in a gas phase, a_{wv}. If water vapour behaved as an ideal gas, then

$$a_{wv} = \frac{P_{wv^*}}{P_{wv}} = \frac{\%\ \text{relative humidity}}{100}$$

where P_{wv} is the partial pressure of water vapour in the air under consideration, and P_{wv^*} is the saturation vapour pressure in equilibrium with pure liquid water at the same temperature. To handle deviations from ideality. P_{wv} is replaced $\gamma_{wv}P_{wv}$, where γ_{wv} is the activity coefficient (more properly, the "fugacity" coefficient) of water vapour. If water vapour obeyed the idea gas law—$P_jV=n_jRT$, where P_j is the partial pressure of species j in a volume V that contain n_j moles of that gas—γ_{wv} would equal unity. This is an extremely good approximation for most situations of interest in plant physiology.

The relative humidity defined by Equation is a readily measured quantity reflecting the activity of water in a particular gas phase, and it has been extensively used in studying the water relations of plants. For water existing as water vapour in air, the chemical potential of the water as defined by Equation is simply μ_{wv^*} + RT ln a_{wv} + m_{wv}gh. Using the expression for the activity of water vapour given by Equation, the water potential of water vapour in a gas phase such as air, ψ_{wv}, is then

$$\psi_{wv} = \frac{\mu_{wv} - \mu_{wv^*}}{V_w} \frac{RT}{V_w} \ln\left(\frac{\%\ \text{relative humidity}}{100}\right) + \rho wgh$$

were a definition of water potential analogous to that given in Equation $[\psi = (\mu_w - \mu_{w^*})/V_w]$ has been used, and m_{wv}/V_w (the same as m_w/V_m has been replaced by the density of water, ρ_w. Note that V_w, not V_{wv}) is used in the definition of ψ_{wv} in Equation. This is necessary since the fundamental term representing free energy per mole is the chemicai potential—we want to compare $\mu_j - \mu_j$ for the two phases when predicting changes at an interface—and thus the proportionality factor between $\mu_w - \mu_{w^*}$ or $\mu_{wv} - \mu_{wv^*}$ and the more convenient terms, ψ or ψ_{wx}, must be the same in each case, viz. V_w.

Let us next consider the various parameters at the surface of pure water (a_w=1) at atmospheric pressure (P = 0 bars) in equilibrium with its vapour. In such a case, P_{wv} equals P_{wv^*}, and we can take the zero level of the gravitational term as the surface of the liquid. Hence μ_w equals μ_w*, and μ_{wv}*. Since there is no tendency for a net gain or loss of water molecules from the liquid phase at equilibrium ($\mu_w = \mu_{wv}$), we conclude that the constant μ_w* must equal the constant μ_{wv}* when the two phases are at the same temperature. What happens to P_{wv} as we move vertically upward from pure water at atmospheric pressure in equilibrium with water vapour in the air?

If we let *h* be zero at the surface of the water, ψ equal zero ($\psi = P - \pi + \rho_w gh$). Equation, and since water vapour is by supposition in equilibrium with the liquid phase, ψ_{wv} is also zero. Equation indicates that as we go vertically upward in the gas phase, $\rho_w gh$ must make an increasingly positive contribution to ψ_{wv}. Since we are at equilibrium, the other term in ψ_{wv}, which equals $(RT/V_w)\ \ln (P_{wv}/P_{wv}*)$ by Equation must make a compensating negative contribution. Thus, at equilibrium P_{wv} must decrease with altitude.

We can further appreciate this conclusion simply by noting that gravity attracts the molecules of water vapour and other gases towards the earth, i.e. air pressure is higher at sea level than it is on the top of a mountain. This effect on P_{wv} is generally not very important in plant physiology, since rather great heights are needed before the partial pressure of water vapour is reduced significantly. In particular, let us consider a 7% reduction in P_{wv} from P_{wv}* (its value at the water surface at equilibrium) to 0.93 P_{wv}*). Since RT/V_w is 1350 bars at 20° (RT/V_w) would change by (1350 bars) in (0.93), or—99 bars pwg equals 0.098 bar/m, hence a height of (99 bars)/(0.098 bar/m), or 1010 m would be required for the gravitation term to increase by the same account. In other words, not until we reached a height of 1 km would the partial pressure of water vapour at equilibrium decrease by 7% from its value at sea level, indicating that the gravitational term generally has relatively little influence on ψ_{wv} over the distance involved for a particular plant.

Plant-air interface

Water equilibrium across the plant-air interface occurs when the water potential in the leaf cells equals that of the surrounding atmosphere. (This presupposes that the leaf and the air are at the same temperature. To measure ψ^{leaf}, the leaf can be placed in a closed chamber and the relative humidity adjusted so that there is not net water gain or loss by the leaf. Such a determination is rather demanding, since small changes in relative humidity have large effects on ψ_{wv}, as we will now show. Extremely large negative values are possible for ψ_{wv}^{air}. In particular, RT/V_w at 20° is 1350 bars. In the expression for ψ_{wv} (Equation this relatively large factor multiplies In (% relative humidity/100), and a wide range of relative humidities can occur in the air. By Equation and ignoring $\rho_w gh$, relative humidity of 100% corresponds to a water potential in the vapour phase of (1350 bars) [In 100/100)] which equals 0 bars (In l=0). This ψ_{wv}^{air} would be in equilibrium with pure water at atmospheric pressure, which also has a water potential of 0 bars.

For a relative humidity of 99%, ψ_{wv}^{air} given by equation equals (1350) [In (99/100)], which is—13.5 bars—remember that In (1–x) is about –x for | x | « 1, such as is the value 0.01 in the present case. Hence, going from 100% to 99% relative humidity corresponds to a decrease in water potential of 13.5 bars. Small changes in relative humidity indeed reflect marked differences in the water potential in air! Finally, we note that a relative humidity of 50% at 20° leads to a ψ_{wv}^{air} of (1350) In (50/100), or –940 bars. One important consequence of the large negative a values of ψ_{wv}^{air} is easily appreciated by considering values of the water potential that may occur in plant leaves.

The actual value of ψ_{wv}^{leaf} depends on the ambient conditions as well as the plant type and its physiological status. The range of values for most mesophytes is –3 to –30 bars, with –5 bars being typical for leaves of a garden vegetable such as lettuce. From equation, the relative humidity corresponding to a water potential of –5 bars is 99.6%. Even during a rainstorm the relative humidity of the air rarely exceeds 99%. Since relative humidity is lower than 99.6% under most natural conditions, water is continually being lost from a leaf having an internal water potential of –5 bars. (Again, it should be emphasized that throughout this discussion the leaf is assumed to be at the same temperature as the air. A few desert plants have a ψ^{leaf} as low as –50 bars. Even in this case of adaptation to arid climates, however, water still tends to leave the plant unless the relative humidity is above 96%.

Pressure in the cell wall water

Plant cells come into contact with air at the cell walls bounding the intercellular air spaces. Thus, the water potential in the cell walls must be considered with respect to ψ_{wv} in the adjacent gas phases. The main contributor to ψ in cell wall water is often the negative hydrostatic pressure arising from surface tension effects at the numerous air liquid interfaces of the cell wall interstices. In turn, $P^{cell\ wall}$ can be related to the geometry of the cell wall pores and the contact angles that occur. The magnitude of the negative hydrostatic pressure that develops in the cell wall water can be analyzed by considering the pressure that occurs in a liquid within a cylindrical pore—the arguments being basically the same as the one presented earlier in this chapter in discussing capillary rise. One of the forces acting on the fluid in a narrow pore of radius r is the result of surface tension, σ. This force equals $2\pi r\sigma \cos \alpha$, where α is the contact angle, a quantity which can be essentially zero for wettable walls.

Since the adhesive forces at the wall are transmitted to the body of the fluid by means of its cohesion, a tension or negative hydrostatic pressure develops in the fluid. The total force resisting the surface adhesion can be regarded as this tension times the area over which it acts, πr^2, hence the force is $\pi r^2 \times (-P)$. Equating the two forces gives the following expression for the pressure developed in the fluid contained within a cylindrical pore:

$$P = -\frac{2\sigma \cos \alpha}{r}$$

Adhesion of water at interfaces generally creates negative hydrostatic pressures (such as that described by Equation in the bulk of the fluid. Such a negative P arising from interfacial interactions has sometimes been treated in plant physiology as a positive matric pressure, a convention that we mentioned earlier.

The strong water-wall adhesive forces, which are transmitted throughout the interfibrillar spaces by water-water hydrogen bonding, conceivably can greatly reduce the water potential in the cell wall, as the following calculation indicates. At 20° the surface tension of water is 72.8 dynes/cm or 7.28×10^{-5} bar-cm. The voids between the fibers in the cell wall are often about 10^{-6} cm (10 nm or 100Å) across; for wettable walls, cos α can equal one. For water in such cylindrical pores, Equation indicates that P would be $-(2)(7.28 \times 10^{-5}$ bar-cm$)(1)/(0.5 \times 10^{-6}$ cm), or –290 bars. This is an estimate of the negative hydrostatic pressure or tension that could develop in the aqueous solution within cell wall interstices of typical dimensions, supporting the contention that $\psi^{\text{cell wall}}$ can be markedly less than zero.

Moreover, the previous discussion of the tensile strength of water indicates that the hydrogen bonding in water can withstand such a tension. We shall now consider the actual pressures that develop in the cell wall water. The relative value of the water potential in the cell wall compared with that in the adjacent vapour phase is of considerable importance in transpiration and for water loss from plant material in general. From Equation ψ_{wv}^{air} corresponding to a relative humidity of 95% at 20° is (1350) ln (0.95), or—69 bars. According to Equation a P of –69 bars would occur in a water-coating cylindrical pore with wettable walls having radius of $-(2)(7.28 \times 10^{-5})(1)/(-69)$, or 2.1×10^{-6} cm, which is 21 nm or 210Å. In other words, water in 21 nm radius interstices in the cell walls of leaf mesophyll cells would be in equilibrium with air of 95%

relative humidity in the intercellular spaces. If the relative humidity near the cell wall surface were decreased, water in such interstices could be lost, but it would still remain in the finer pores where a more negative P or larger tension can be created.

Hence, if the plant material is exposed to air of moderate to low relative humidity—such as can happen for an herbarium specimen—then larger and larger tensions resulting from interfacial interactions develop in the cell walls as the plant material dries out. Water is then retained only in finer and finer pores of the wall, as it must, according to Equation. We shall now show the effect of the availability of water adjacent to that in wettable cell walls on the contact angle and on $P^{cell\ wall}$. Suppose that pure water in mesophyll cell wall interstices 10 nm across is in equilibrium with water vapour in the intercellular air spaces where the relative humidity is 99% ψ_{wv} for this relative humidity is –13.5 bars at 20°. Hence, at equilibrium the (pure) water in the cell wall interstices must be under a hydrostatic pressure of –13.5 bars ($\psi = P - \pi + \rho_w gh$, Equation. From Equation ($P = -2\ \sigma \cos \alpha / r$), and letting σ be 7.28×10^{-5} bar-cm at 20°, we calculate that $\cos \alpha$ must be – (–13.5 bars) (0.5×10^{-6} cm)/[(2)(7.28×10^{-5} bar-cm)], or 0.0464 for the cell wall pores which are 5 nm in radius.

When $\cos \alpha$ is 0.0464 , α is 87°. For the wettable cell walls, α can be 0°, in which case $\cos \alpha$ is unity and the maximum negative pressure does develop, viz., —290 bars for the pores 5 nm in radius. On the other hand, when water is available, it can be pulled into the interstices by such possible tensions and there by cause α to increase toward 90°, which leads to a decrease in cos a and consequently a decrease in the absolute value of the pressure developed. In the present example, α is 87°, P is –13.5 bars, and the pore 5 nm in radius is nearly filled. Thus, the water in the interstices can be nearly flush with the cell wall surface.

In fact, the large tensions that could be present in the cell wall generally do not occur in living cells, since water is usually available within the plant and is "pulled" into the interstices, thus nearly filling them, depending of course on their dimensions Equations. The only contribution to the water potential of the cell wall that we have been considering is the negative pressure that develops because of air-liquid interfaces, which can be the case for pure water. When the gravitational term in Equation 35 is ignores, $\psi^{cell\ wall}$ actually equals $P^{cell\ wall} - \pi^{cell\ wall}$ The bulk solution in the cell wall pores generally has an average osmotic pressure of 3 to 15 bars. The local $\pi^{cell\ wall}$ next to Donnan phase is considerably greater because of the many ions present there by way of compensation, $P^{cell\ wall}$ is also higher near the Donnan phase than in the bulk of the cell wall water.

Water flux

When the water potential inside a cell differs from that outside, the water is no longer in equilibrium and we can expect a water movement toward the region of lower water potential. Our attention here will be specifically focused on water flow into and out of plant cells. This volume flux of water, J_{vw}, is assumed to be proportional to the difference in water potential ($\Delta\psi$) across the membrane or membranes restricting the flow. The proportionality factor indicating the permeability to water flow at the cellular level is expressed by a water conductivity coefficient, L_w :

$$J_{vw} = L_w \Delta\psi = L_w(\psi^o - \psi^i)$$

In Equation J_{vw} is the volume flow of water per unit area of the barrier per unit time. It can have units of $cm^3/(cm^2 - sec)$ or cm/sec. Which is a velocity. In fact J_{vw} is the average velocity of water moving across the barrier being considered. L_w can be given in cm/(sec-bar), in which case the water potentials would be expressed in bars. (L_w is usually experimentally the same as L_p, a coefficient also describing water conductivity.

When Equation is applied to cells, ψ^o is the water potential in the external solution, and ψ^i usually refers to the water potential in the vacuole. L_w then reflects the conductivity for water flow across the cell wall, the plasmalemma, and the tonoplast, all in series. For a group of barriers in series with an overall water conductivity coefficient L_w, $1/L_w$ equals

$$\sum_j 1/L_{wj}$$

where L_{wj} is the water conductivity coefficient of barrier j. To see why this is so, let us go back to Equation, $J_{vw} = L_w\Delta\psi$. When the barriers are in series, J_{vw} is the same across each one, and so J_{vw} equal $L_{wj}\Delta\psi_j$ (also $\Delta\psi_j = J_{vw}/L_{wj}$) where $\Delta\psi_j$ is the drop in water potential across barrier *j*, i.e.,

$$\sum_j \Delta\psi_j = \Delta\psi.$$

We therefore obtain the following string of equalities:

$$J_{vw}/L_w = \Delta\psi = \sum_j \Delta\psi_j = \sum_j J_{vw}/L_{wj} = J_{vw}$$

$$\sum_j 1/L_{wj};$$

hence, $1/L_w$ must equal

$$\sum_j 1/L_{wj}.$$

We also note that $1/L_w$ corresponds to a resistance and the resistance of a group of resistors in series is the sum of resistances,

$$\sum_j 1/L_{wj}.$$

We shall now consider a membrane separating two solutions that differ only in osmotic pressure. This will help us to relate L_w to the permeability coefficient of water, P_w, and also to view Fick's first law [$J_j = P_j(c_j^{\circ} - c_j^{\circ})$, in a slightly different way. The proper form for Fick's first law when describing the diffusion of the solvent water is $J_{vw} = P_w(N_w^{\circ} - N_w^{i})$. By Equation, J_{vw} also equals $L_w(\psi^{\circ} - \psi^{i})$, which becomes $L_w(\pi^{i} - \pi^{\circ})$ in our case (recall that $\psi = P - \pi + \rho_w gh$, Equation. Thus, a flux of water toward a region of higher osmotic pressure (Equation) is equivalent to J_{vw} toward a region where water is less concentrated in the sense of having a lower mole fraction (Fick's first law applied to water). Let us pursue this one step further. Using Equations through 40 we can obtain the following string of equalities:

$$N_W \cong 1 - \sum_j n_j / nw \cong 1 - \overline{V}w\pi_j / RT.$$

Therefore, $N_w^{\circ} - N_w^{i}$ equal $(1 - V_w\pi_s^{\circ}/RT) - (1 - V_w\pi_2^{i}/RT)$, or $V_w(\pi_s^{i} - \pi_s^{\circ})/RT$. When we incorporate this last relation into our two expressions for the volume flux of water, we obtain $J_{vw} = L_w(\pi^{i} - \pi^{\circ}) = P^{w}(N_w^{\circ} - N_w^{i}) = P_w V_w(\pi_s^{i} - \pi_s^{\circ})/RT$ (π^{i} is here the same as π_s^{i}). Here, L_w equals $P_w V_w/RT$, which states that the water conductivity coefficient (L_w) is proportional to the water permeability coefficient (P_w).

Next, we will estimate a possible value for L_w. P_w is generally at most 10^{-2} cm/sec for plasmalemmas, RT is 24,370 cm^3bar/mole at 20° and V_w is 18 cm^3/mole. Hence, L_w might be (10^{-2} cm/sec) (18 cm^2/mole)/(24,370 cm^3-bars/mole) or 0.74×10^{-5} cm/sec-bar). When P_w is less than 10^{-2} cm/sec, L_w will be proportionally less than 0.74×10^{-5} cm/sec-bar). When the value of the water conductivity coefficient is known, the water potential difference necessary to give an observed water flux can be calculated by using Equation. For the internodal cells of *Nitella* and *Chara*, L_w for water entry can be about 10^{-5} cm/(sec-bar)—a large value for L_w, see Dainty. For convenience of calculation, we will consider cylindrical cells 10 cm long and 1 mm in diameter as an approximate model for such algal cells. The surface area across which the water flux occurs is $(2\pi r)(l)$, where r is the cell radius (0.05 cm) and l is the cell length (10 cm). Thus, the area equals $(2\pi)(0.05\text{ cm})$, or $\pi\text{ cm}^2$. (The area of each end of the cylinder (πr^2) is much less than $2\pi rl$ in any case, a water flux from the external solution across them would not be expected, since they are in contact with other cells, not the bathing solution. The volume of a cylinder is $\pi r^2 l$, which equals $(\pi)(0.05\text{ cm})^2$ (10 cm), or $\pi/40\text{ cm}^2$ in the present case.

Internodal cells of *Nitella* and *chara* grow relatively slowly—a change in volume of about 1% per day being a possible growth rate for the fairly mature cells which were used in determining L_w. This growth rate means that the water content increases by about 1% of the volume per day. The total volume flux of water into the above cell is therefore (0.01) $\pi/40$) cm^3/day. Such an inflow of water per day is equivalent to [(p/4000) cm3/day] (1/24 day/hr) (1/3600 hr/sec) or $3\pi \times 10^{-0}\text{ cm}^3$ of water moving into a cell/sec. This water influx occurs across a surface area of $\pi\text{ cm}^2$ per cell, as we just calculated, hence the rate of volume flow of water per cm^2 is $(3\pi \times 10^{-9}\text{ cm}^2/\text{sec})/(\pi\text{ cm}^2)$, or $3 \times 10^{-9}\text{ cm}^3/(\text{cm}^2\text{ sec})$. But such a quantity is precisely what is meant by J_{vw} in Equation.

Using the appropriate value of L_w for these internodal cells, 10^{-5} cm/sec-bar, Equation indicates that $\Delta\psi$ is $(3 \times 10^{-9}\text{ cm/sec})$ (10–5 cm/sec-bar) or 3×10^{-4} bar. In order words, the internal water potential (ψ^i) needed to sustain the water influx accompanying a growth of about 1% per day is 3×10^{-4} bar less than the outside water potential (ψ^o). *Nitella* and *Chara* can grow in pond water, which is a dilute aqueous solution often having a water potential near –0.07 bar. Thus, ψ^i need be only slightly more negative than –0.07 bar to account for the influx of water accompanying a growth of 1% per day.

4

SOLUTES IN PLANTS

From the time that a young plant starts to grow until its death, a more or less continuous movement of solutes is in progress through the conducting elements of every organ of the plant. In a very young seedling, foods are usually translocated upward in the growing stems and downward in the developing roots from the storage tissues of the seed. As soon as the rate of photosynthesis in the developing seedling becomes sufficiently high, at least part of the photosynthate moves in a downward direction from the leaves toward the roots while some of it is often moving upward toward apical meristems. Furthermore, as soon as the developing roots make effective contact with the substrate, absorption of mineral salts begins, followed by translocation of a large part of them in a generally upward direction through the plant. At least some of the solutes absorbed from the soil, especially those containing nitrogen, phosphorus, or sulphur, often react within the root cells with organic compounds which have descended into the roots from the leaves.

The resulting chemically more complex compounds, such as amino acids and acid amides, may then he translocated in the reverse direction from the roots into the aerial parts of the plant. Mineral salts absorbed by the roots are mostly translocated to young leaves and other growing organs of the plant. All of them do not remain, however, in the organ into which they are first translocated. A considerable proportion of the mineral salts which move into a leaf or a flower petal, for example may sooner or later move out of such a lateral organ back into the stem, and become redistributed to other usually younger parts of the plant. The translocation patterns within a plan often exert significant effects upon its behaviour. At the time that young fruits and seeds are developing, for example, there appears to be a general migration of organic materials from all parts of the plant toward the enlarging fruits and seeds. This movement may so nearly monopolize the food resources of the plant as to check severely the growth of vegetative organs. In cotton plants, for example, there is a conspicuous decrease in vegetative growth when the plant is fruiting heavily.

A large part of the carbohydrate synthesized in photosynthesis and of the nitrogenous compounds in the above ground organs of the plant moves into the enlarging fruits. The supply of carbohydrates reaching the root system becomes insufficient to maintain rapid respiration and root elongation. As a result here is a decrease in the rate of absorption of mineral salts, which in turn results in checking or even stopping the vegetative growth of the plant. The patterns of translocation in plants are complex and may be different at different stages in the life history of the plant. Many kinds of solutes are being translocated in various directions within the plant. Nevertheless certain predominant translocation routes can be recognized as follows: (1) downward translocation of organic solutes from leaves to other parts of the plant, (2) upward translocation of organic solutes to growing or storage regions, (3) upward translocation of mineral salts from roots to aerial organs, (4) outward translocation of mineral salts from leaves and other lateral organs into stems. (5) lateral or cross transfer of solutes within stems.

Downward Translocation of Organic Solutes

Downward translocation of organic compounds unquestionably occurs for the most part through the phloem tissues. Much of the evidence indicating that organic solutes move toward the basal portions of the plant in the phloem has been obtained by ringing experiments. "Ringing," when the term is employed without qualification, refers to the removal of a narrow continuous band of tissues external to the xylem. Since ringing entirely encircles the stem, all tissues external to the xylem are completely intercepted. This operation is also called "girdling." In a girdled tree carbohydrates and other organic compounds slowly accumulate in the tissues above the ring and slowly decrease on quantity in the tissues below the ring as they are utilized in respiration and assimilation.

Unless special conditions intervene, such, for example, as development of sprouts on the tree trunk below the ring, a girdled tree ultimately dies because of starvation of the roots, showing clearly that no appreciable amounts of foods are conducted downward through the xylem. Accumulation of organic compounds above a ring has also been demonstrated in certain herbaceous plants such as cotton. Experimental results of this kind indicate that the downward translocation of carbohydrates and other organic compounds occurs through the phloem tissues. Chemical analyses show that cells of the phloem are relatively rich in carbohydrates and organic nitrogenous compounds. This finding is consistent with the concept that translocation of organic compounds occurs through the phloem but is not proof, since storage tissues also contain relatively high concentrations of foods. We know that the upward movement of water takes place in the xylem.

For this reason it is difficult to conceive of the xylem as an important avenue of downward translocation of organic substances. Dyes injected into stems at times of rapid water movement usually move both upward and downward sometimes to approximately equal distances. Such results have sometimes been cited as evidence of downward currents in the xylem. Downward movements of injected dyes probably result, however, from the effects of tension or from a subatmospheric pressure in vessels occupied only by gasses. They cannot be accepted as valid evidence that downward movements of solutes ordinarily occur in the xylem of intact stems. Further evidence that the phloem is the tissue in which organic compounds are translocated has come from experiments in which short segments of the stem have been killed. Since the conducting elements of the xylem are nonliving, the passage of solutes through the xylem is not prevented by the death of the cells in the stem.

The conducting elements of the phloem, on the other hand, are living hence any phase of translocation which is interrupted be killing the cells in a stem segment can be assumed to be occurring in the phloem. Rebideau and Burr (1945) labeled the carbohydrate formed in bean plants by allowing one leaf on the plant to photosynthesize in an atmosphere containing carbon dioxide made with the identifiable C^{13} isotope. They then showed that movement of carbohydrates occurred readily in the stem, both upward and downward, from the node at which the leaf petiole was attached. If short segments of the steam above and below the node at which the petiole was attached were killed by treatment with hot wax, however, no upward or downward movement of carbohydrates occurred. Phosphates, containing radioactive phosphorus (P^{32},) however, were readily conducted through such killed stem segments, indicating that their translocation of mineral salts.) Bonner (1944) and Went (1944) obtained similar results in investigations of the translocation of organic compounds through tomato stems, short segments of which had been killed with superheated steam.

Upward Translocation of organic Solutes

Under many conditions and upward translocation of organic solutes takes place in plants. This occurs, for example, in the stems of woody species when the buds resume growth in the stems of woody species the direction of translocation of carbohydrates ; they are then translocated from the leaves into the stems in which they move to other organs of the plant. A number of other examples of the upward transport of foods in plants can be cited. Opening flowers and developing fruits are often attached to stems in such a position that some or all of the organic compounds translocated into them move through the stems in an upward direction. In the early stages of the development of seedlings, upward translocation occurs from the endosperm or cotyledons toward the apical portions of the plant in which rapid growth is taking place.

Likewise upward transport of foods invariably occurs during the earlier stages of shoot growth from bulbs, tubers, rhizomes, and other types of underground organs. The "classical" view that upward translocation in plants takes place in the xylem was long accepted as referring to mineral salts and organic compounds as well as to water. The concept that upward transport of carbohydrates occurs principally through the xylem is based largely on phenomena which have been observed in certain woody species. Large quantities of soluble and insoluble carbohydrates are present in the wood parenchyma, wood rays, and (in the younger stems) pith cells of many varieties of trees and shrubs.

At certain seasons soluble carbohydrates are also found in the xylem conduits, as illustrated by data of Anderson (1929). As found by this investigator the concentrations of both sucrose and free reducing substances (probably largely hexoses) in the xylem sap of pear trees were highest in the winter and early spring. Both fell to a zero value during the summer months, and increased slowly during the autumn. The sugars found in the conducting elements undoubtedly come from the storage tissues of the pith or xylem. The relatively high soluble carbohydrate content of the xylem tissues in the winter and spring results largely from shifts in the starch sugar equilibrium toward the sugar side, as a result of the relatively low temperatures prevailing during these seasons. The fact that, in the early spring, the soluble carbohydrate content of the xylem sap is relatively high and by the time the shoots are well developed has dropped to a low or zero value seems to indicate that soluble carbohydrates have been conducted upward through the xylem to the developing buds. Although this assumption appears to be superficially

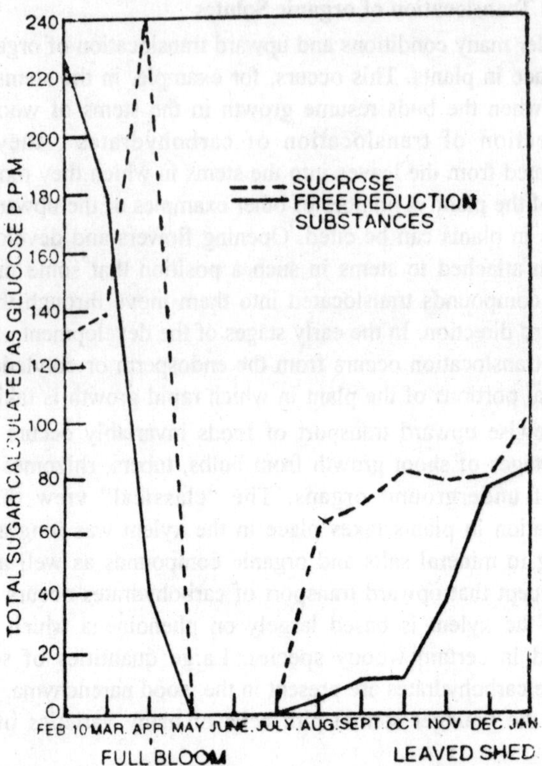

Fig. 4.1. Seasonal variation in the carbohydrates content of the xylem sap of pea trees.

plausible, there are good reasons for doubting if this phenomenon actually is very strong evidence that upward translocation of organic solutes occurs in the xylem.

In the first place the concentration of organic solutes in the xylem sap is always very low, seldom exceeding 2 to 3 per cent. Furthermore, the highest concentrations of sugar occur in the sap during the winter when there is little or no upward translocation of water, and it is not at all certain that these solutes do not largely disappear from the xylem sap in the spring before its upward flow begins to take place at an appreciable rate. Investigations by Curtis, summarized in 1935, point to a conclusion regarding the upward transport of organic solutes which is directly the opposite of the long accepted view. His experiments all seem to indicate that it is the phloem and not the xylem which is the principal tissue through which such translocation occurs.

In case investigation the contrasting effects of intercepting the xylem and intercepting the phloem of woody stems upon upward transport of organic solutes were studied (Curtis, 1925). In these experiments a number of growing shoots were first defoliated. Some received no further treatment, thus serving as checks. In others a ring of the tissues external to the xylem was removed, and in still others a segment of the xylem was excised, leaving the phloem and cortical tissues intact. Every stem which was cut into was enclosed in a glass cylinder as shown in the figures. This cylinder was filled with water in order to keep the exposed tissue surfaces moist and in order to supply water to the top of the stems in which the xylem was cut.

The outcome of the experiments in which the water jacket was rinsed out once each day with distilled water was essentially the same as in those experiments in which this was not done, indicating that translocation of solutes did not occur through the water. In variably the stems in which the xylem was cut showed greater elongation than those in which the phloem was cut, indicating greater upward translocation of foods in the former than in the latter. Shoot elongation is a somewhat indirect measure of translocation, but the conclusion drawn from such observations were supported in a number of experiments by dry weight determinations and analyses for sugar.

The dry weight and total sugar content per stem were invariably lease in the ringed stems. This was also true for the percentage of sugar in terms of fresh weight or dry weight. In general the evidence seems to warrant the conclusion that the phloem is the principal tissue in which upward translocation of organic solutes occurs, although it seems probable that small quantities of such compounds are, on occasion, translocated in an upward direction through the xylem.

Upward Translocation of Mineral Salts

The term "mineral salts" is necessarily employed in this discussion somewhat loosely. For present purposes, nitrogen, which may be translocated upward in either inorganic or organic combination is included among mineral salts. Upward translocation of sulphur and phosphorus also probably occurs in both organic and inorganic combination. For many years it was universally agreed that upward translocation of mineral salts occurred through the xylem although it now appears that the situation is not quite this simple. Studies of the sap from xylem vessels show that it usually contains at least traces of both organic and inorganic solutes.

In proportion to the total quantities utilized, however, the concentration of inorganic constituents in the xylem sap is usually relatively higher than that of organic solutes. Further more, appreciable concentrations of mineral salts are commonly present in the sap of vessels at season when upward flow of water is occurring at its most rapid rates. At such times the xylem sap contains little or no organic material in solution. Presence of dissolved; mineral salts at such seasons is presumptive evidence that at least some of them are translocated upward in the plant through the xylem. Clements and Engard (1938), Phillis and Mason (1940), and others have shown that ringing stems of various species does not prevent upward movement of mineral salts through the plant. Results of such experiments show that upward translocation of mineral salts can occur in the xylem.

Curtis employed the technique of intercepting the xylem versus intercepting the phloem in attempting to ascertain whether mineral salts moved into growing, defoliated stems of sumac through the xylem or through the phloem. Relative to the quantity which moved through check stems, a much larger proportion of mineral salts or nitrogenous compounds was translocated through the stems in which the xylem was intercepted than through those in which the phloem was intercepted. Movement of mineral salts into such woody shoots can therefore occur in the phloem. Somewhat similar experiments performed by Mason and Phillis (1940) on cotton plants also led to the conclusion that upward translocation of nitrogenous compounds can occur in the phloem. Experiments on ringing often show that mineral salts can be conducted in the xylem, and experiments on intercepting would often show that they can be conducted in the phloem, but no such experiments show conclusively which is the predominant pathway of such conduction in intact plants.

Stout and Hoagland (1939) were among the first to employ the radioactive tracer technique in experiments designed to ascertain the path of upward movement of mineral salts in plants. Small plants of cotton, geranium, and willow, rooted in sand or solution cultures, were used. Certain branches of each plant were "stripped" by cutting longitudinal slits 9 in. long on opposite sides of the stem, and then carefully pulling the bark away from the wood, but leaving it attached at the ends. A sheet of paraffined paper was then introduced between the phloem and the xylem. This treatment resulted in no visible signs of injury to the plants during the course of an experiment. Radioactive ions of potassium, sodium, phosphate, or bromide were introduced into the rooting medium.

After a period of not more than a few hours, under conditions favourable to transpiration, distribution of the tracer ions in the stem

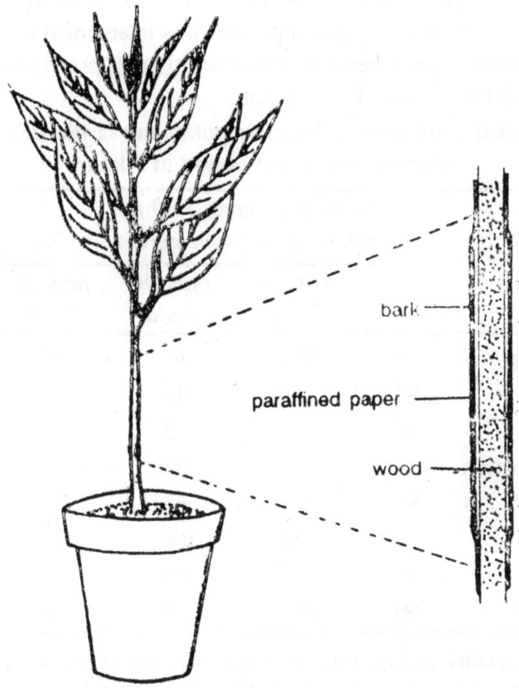

Fig. 4.2. Diagram to ılustrate technique employed in the experiment of Stout and Hoagland.

was ascertained by measuring the quantity of radioactive elements in a shed segments of the xylem and phloem from the stem above, below, and in the region where the xylem and phloem were separated with paraffined paper. As shown by the results from a representative one of their experiments, radioactive potassium (K^{42}) was found to be relatively abundant in both the bark and the wood above and below the section of the willow stem in which the xylem and phloem had been separated by paraffined paper. Within this latter segment of the stem, however, almost all of the tracer element was located in the xylem. In the intact branch there was no such marked difference in the distribution of the radioactive potassium.

Similar results were obtained with the other plants used, and with each of the radioactive elements employed as a tracer. In these experiments it is obvious it is obvious that mineral salts absorbed by the roots were transported upward at a relatively rapid rate through the xylem. During their upward passage some of the mineral salts moved laterally from the xylem into the phloem, thus accounting for the relatively large amounts of tracer element in the phloem. When, however, the lateral movement of salts from xylem to phloem was intercepted by interposing an impermeable barrier between these two tissues, practically all of the tracer element was found in the xylem.

Table 4.1. Distribution of radioactive potassium in the stem of willow after as absorption period of 5 hours

		Branch stripped 1.30 hr before absorption period		*Intact branch*	
		ppm in bark	*ppm in wood*	*ppm in bark*	*ppm in wood*
Above strip	SA	53	47	64	56
	S6	11.6	119		
	S5	0.9	122		
Stripped section	S4	0.7	142	87	69
	S3	<0.3	68		
	S2	<0.3	108		
	S1	20	103		
Below strip	SB	84	58	74	67

These results demonstrate that upward movement of mineral salts occurs in the xylem, but the possibility of some such movement in the phloem also is not excluded even though it is clear that lateral movement from xylem to phloem takes place readily. The data furnish some evidence of upward and downward translocation in the phloem. Bark sections S1 and S6 adjacent to the unstripped portions of the stem, are both higher in radioactive potassium that sections S2 to S5. This can be accounted for only by assuming that a slow movement of radioactive potassium has occurred into these sections from adjacent unstripped portions of the bark. Somewhat similar experiments of Gustafson (1939) also indicate that a limited amount of upward translocation of radioactive phosphate may occur through the phloem. There is no doubt that upward translocation of mineral salts occurs in the xylem, and it seems virtually certain that this is the main pathway along which their general upward movement from roots to leaves occurs.

Conditions which are especially favourable to mineral salt translocation in the xylem include high rates of transpiration in the xylem include high rates of transpiration, high concentrations of mineral salts in the substrate, and the prevalence in the root cells of metabolic conditions which favour rapid movement of mineral salts from the absorbing cells to the xylem. Some upward translocation of mineral salts also occurs in the phloem under certain circumstances. Such movement appears to occur at rates that are usually slower than in the xylem. As described in the next section, upward movement of mineral salt in the phloem of younger stems appears to be a usual occurrence when they are "exported " from leaves into the stem.

There are some indications that mineral elements are more likely to move in the phloem when in organic than when in inorganic combination. Entrance of such elements as nitrogen, phosphorus, and sulphur into organic combinations in the root cells may favour their upward translocation in the phloem.

Outward movement of solutes other than carbohydrates from leaves and other lateral organs

As the previous discussion has indicated, movement of minerals salts to the leaves appears to occur principally, if not wholly, in the xylem. Not all of the mineral salts which enter a given leaf remain in that leaf, however, some of them being "exported" back into the stem, whence they are translocated to other parts of the plant. This is shown by the results of chemical analyses of leaves at different times of the day and also by results of periodic chemical analyses of leaf and stem tissues during the period just prior to leaf abscission. Nitrogen, phosphorus, potassium, sulphur, magnesium, and chlorine may all be exported from leaves in one form or another, but calcium, boron, iron, and manganese appear to be virtually immobile.

Nitrogen, phosphorus, and sulphur probably move out of leaves at least partly in organic combination, while potassium, magnesium, and chlorine probably move out mostly in the form of inorganic ions. Translocations of mineral elements out of flower petals just prior to their abscission also occurs. In general, those mineral elements which are readily redistributed in plants are the ones which are the most readily, exported from leaves. The earlier evidence of the export substances other than carbohydrates from leaves and other lateral organs has been augmented and strengthened by more recent studies in which radioactive elements have been used. To study this phenomenon by means of tracer elements it is necessary to apply the tagged ions directly to the leaf tissues rather than to let them enter the plant in the usual way through the root system.

The manner in which the radioactive elements are introduced into the leaf tissues may determine the channel through which they pass in moving from the leaf to the stem. If the solution containing the tagged compound is brought into contact with the open ends of several xylem vessels it may be drawn rapidly into the stem by way of the xylem. This is particularly likely to happen if an internal water deficit exists in the leaf at the time the solution is applied. On the other hand, when the marked ions are allowed to reach the vascular system by cell-to-cell transfer in the mesophyll, the result obtained is very different. Colwell (1942) found that radioactive phosphorus (P^{32}), introduced as phosphate into intact leaves of squash plants under suitable conditions, moved out of such leaves into the stem readily but did not move out of similar leaves if the petioles had first been sealded. This is evidence that outward movement of the phosphorus-containing compounds occurred in the phloem. Biddulph and Markle (1944) introduced radioactive phosphorus as phosphate into the leaves of cotton plants in such a way as to insure that the tagged ions would move into the phloem tissue of the leaf.

The xylem and phloem of the stem were separated by a membrane of waxed paper in a segment extending for 4 in. immediately below the point of leaf attachment. The phosphate ions were allowed to migrate through the tissues of the leaf and stem for from 1 to 3 hr. before their distribution in the stem was determined. Examination of the treated portion of the stem revealed that almost all of the radioactive phosphate was located in the bark. The phosphate had moved both up and down after reaching the stem, but only traces of the tagged element were found in the xylem where it was separated from the phloem by the waxed paper. It is evident from these results that the phosphate was translocated out of the leaf and longitudinally within the stem almost entirely in the tissues of the phloem. When the radioactive phosphate was applied to leaves of plants with intact stems, the marked ions were found in both the xylem and the phloem of the stem and both above and below the point of leaf attachment.

Again, much more of the radioactive element was present in the bark than in the wood which suggests that the tracer ions reached the latter by lateral diffusion from the phloem. Studies of the outward movement of radioactive phosphate from the leaves of bean plants indicated that more such movement occurred during the daytime than at night. The direction of the movement of phosphate was principally downward, under the conditions of these experiments, but some lateral diffusion of the radioactive ions into the xylem took place with a consequent upward movement in the transpirations stream. Other evidence shows quite clearly that a diurnal alternation between a movement of mineral salts into leaves and a movement out of the leaves is of frequent occurrence.

Phillis and Mason (1942) have shown by chemical analyses that the magnesium, potassium, chlorine, phosphorus, and nitrogen content of cotton leaves regularly increases during the daylight hours, and decreases at night. The general picture of the movements of mineral salts in plants seems to be about as follows. After entering the roots from the soil most of the mineral salts move across the cells of the young roots to the xylem ducts in which they are carried upward into the leaves. Some of the molecules of which these elements area a part or of which they become a part as a result of chemical transformations in the leaf do not become immobilized in the leaves but move out of them via the phloem. Once in the stem such solutes move both upward and downward in the phloem, entering other leaves, and probably also reaching the apical growing regions of roots and stems. Some of the mineral compounds in the phloem also migrate laterally and enter the transpiration stream in which they are transported upward at a usually rapid rate. In a sense, therefore, we may speak of a "circulation" of mineral compounds within the various organs of a plant.

Lateral Translocation of Solutes

Lateral translocations of water in a tangential direction readily occurs in woody stems. This does not appear to be true of many solutes. In straight-grained trees the sugars from the leaves on one side of the tree are translocated to the roots directly below them and, if nitrates are added to the soil on one side of a tree, increase in nitrogen occurs principally in the leaves and branches above the roots on that side. A similar lack of lateral translocation of solutes has been shown in certain species of herbaceous plants. On the other hand, radial transfer of solutes from xylem to phloem or *vice versa* appears to occur readily. Some of the radial movement of solutes probably occurs along the vascular rays. In perennial woody plants, however, there is evidence that reorientation of conductive tissues may occur in such a way as to offset the effects of a lack of lateral translocation of solutes. If all branches on one side of a tree are removed or destroyed, for example, the effect of diminishing growth of trunk and roots on that side usually does not persist for more than one year. Subsequently developed conductive tissues are usually oriented in such a fashion as to permit translocation to the trunk and roots on the side of the tree which no longer bears branches.

Mechanism of Translocation of Solutes in the Xylem

Dissolved mineral salts and organic solutes, when present, are carried along with the ascending streams of water in the xylem ducts which are pulled up through the plant according to the mechanism. During upward translocation some of the solutes are lost by lateral movement into living cells of the stem adjacent to the xylem conduits. In the leaves the solutes of the xylem vessels migrate into the living cells of the mesophyll. The rates at which solutes are translocated upward through the xylem of the stem will correspond with the rates of translocation of water.

Basic Considerations Regarding Mechanism of Translocation of Solutes in the Phloem

Although several theories have been advanced to account for the mechanism of the movement of solutes in phloem tissues, no one theory has ever received general acceptance. As a background for considering the principal theories that have been proposed, it will be helpful to review briefly some of the pertinent facts regarding the problem of movement of solutes in the phloem.

Living cells are essential

Unlike movement of water and solutes in the xylem, translocation in the xylem, translocation in the conducting elements of the phloem ceases when the cells are killed. Evidence for this fact is presented earlier in the discussion.

The movement may be bidirectional

There is no doubt that movement of solutes in the phloem can sometimes occur in one direction and sometimes in the other. The question of whether or not simultaneous bidirectional movement can occur through the same region of phloem has an important bearing on concepts of the mechanism. Experiments already discussed show that both carbohydrates and phosphates move both up and down in the phloem of a stem from the point at which they enter it from the petiole of a leaf. This phenomenon does not, however, involve bidirectional movement through the same problem at the same time. Indications that two different kinds of solutes—sugar and the dye flurorescein—could move simultaneously through the same phloem region the opposite directions were obtained by Palmquist (1938). Simultaneous movement of phosphate and carbohydrate, labeled with radioactive phosphorus (P^{32}) and radioactive carbon (C^{14}) respectively, in opposite directions through the phloem tissue of geranium stems has been shown to occur, at least over short distances and a relatively slow rates. It was further shown that phosphate could move up in the phloem while carbohydrate was moving down in the same phloem, and *vice versa*.

Chemical Potential of Ions

Since a substance spontaneously tends to move to regions where its chemical potential is lower this quantity is a useful parameter for analyzing passive movements of solutes. Using a linear combination of the various contributors, the chemical potential of any species j is given by Equations as follows : $\mu j = \mu j^* + RT \ln aj + VjP + zjFE + mjgh$. Since water is uncharged ($zw = 0$), the electrical term does not enter into its chemical potential. For ions, however, zjFE becomes important. In fact, for charged solutes under most conditions of biological interest, differences in the electrical term are usually far larger than are changes in the terms for the hydrostatic pressure or gravity. In particular, for movements across membranes, Δh is zero and so we will omit the gravitational term in this chapter, while changes in the VjP term are also generally ignored because they are relatively small (as we will indicate below). Therefore the chemical potential generally used when dealing with ions, $\mu j^* + z_j FE$, is commonly referred to as the *electrochemical* potential. Although this emphasizes the role played by electrical potentials, "chemical potential" is only a shorthand expression for the sum of all the different contributors affecting a particular species, and it is not really necessary to use a special term like "electrochemical" potential when dealing with ions.

Electrical Potential

The difference in electrical potential E between two locations is a measure of the amount of work involved in moving a charge from one location to the other. When the charge is expressed in coulombs and the potential difference in volts, then the amount of work in joules equals the product of the two. The zero level for electrical potential is actually arbitrary. Since both the initial and the final electrical potential must be expressed relative to the same zero level, the arbitrariness is of no consequence when the electrical potential difference (final minus initial potential) is determined. The charge carried by an ion of species *j* is a positive or negative integer zj (the charge number) times the charge on a proton. For instance, zj is +1 for the potassium ion and –2 for sulphate (SO_4^{--}).

The electrical charge carried by proton is commonly called the *electronic charge*, since it is the same in magnitude as the charge on the electron (but opposite in sign, of course). One unit for charge is the coulomb; a proton has a charge of 1.6021×10^{19} coulomb. Thus, a mole (Avegadro's number) of protons would have a charge equal to (6.02252×10^{23} protons/mole) (1.6021×10^{-19} coulomb/proton), or 96,487 coulombs/mole. Such a unit consisting of Avogadro's number of electronic charges (i.e., one mole of single, positive charges) is called the Faraday, F. This quantity appearing in the electrical term of the chemical potential equals

96,487 coulombs/mole or 96,487 joules/mole-volt—a coulomb-volt is a joule—which can also be expressed as 23,060 cal/mole-volt. To illustrate the rather large contribution the electrical term can make to differences in the chemical potential of a charged substance compared with the effect of changes in hydrostatic pressure, let us consider a small monovalent cation (zj = + 1) that has a partial molal volume (Vj) equal to 30 cm^3/mole. What electrical potential difference, (ΔE) in volts (v) corresponds to the same contribution to the chemical potential of such an ion as a hydrostatic pressure drop (ΔP) of one bar ? In other words, what is ΔE so that zjFΔE equals VjΔP under these conditions ? One bar is 10^6 dynes/cm^2, or 0.1 joule/cm^3. Thus, ΔE need be only (30 cm^3/mole) (0.1 joule/cm^3)/[(1)(96,487 joules/mole-volt)] that is, 3.1×10^{-5} v (0.031 mv). This is an extremely small electrical potential difference !

For comparison, the drop in electrical potential across a biological membrane is often near 100 mv, which is (1000)/(0.031) or 3200 times larger than the potential difference calculated here. Stated another way, 100 mv across a membrane would make the same contribution to the chemical potential difference as a drop of 3200 bars of hydrostatic pressure from one side of the barrier to the other for the above case of a monovalent cation having a Vj of 30 cm^2/mole. Most hydrostatic pressures encountered in plant or cell physiology are only a few bars. Thus, hydrostatic pressure drops across membranes are usually negligible in comparison with the electrical contributions to the chemical potential differences of ions from one side of the membranes to the other.

Electroneutrality and Membrane Capacitance

Another consequence of the relatively large magnitude of electrical effects is the general occurrence of *electroneutrality*. Thus, in most aqueous regions that are large compared with atomic dimensions the total electrical charge carried by the cations is essentially equal in magnitude to that carried by the anions. What is the situation for plant cells? Do we ever have an excess of negative or positive charges in cells or organelles? If there were a net charge in some region, such as near a membrane, an electrical potential difference would exist from one part of the region to another. Can we relate the size of the electrical potential drop to the net charge? In order to relate charges and ΔEs, we need to introduce a new term, *capacitance*.

Electrical capacitance is simply a coefficient of proportionality between a net charge and the resulting electrical potential difference. A high capacitance means that the region has the *capacity* to have many uncompensated charges without at the same time developing a large electrical potential drop. Let us now express this symbolically. The magnitude of such an electrical potential difference, ΔE, is related to the capacitance of the region (C) as follows :

$$Q = C\Delta E \qquad ...(1)$$

where Q is the net charge. The unit for capacitance is the *farad* (f), which equals 1 coulomb/volt. The capacitance of most biological membranes are approximately the same per unit are, about 1 microfarad (μf)/cm ^{2} (see Bull. Cole. Keynes in Agin, or Ruch and Patton). For convenience in estimating the net charge inside a cell which would lead to a typical membrane potential, we will consider a spherical cell of radius *r.* Suppose that the uncanceled or net charges are uniformly distributed in space at a concentration, c. (For a conductor—a body in which electrical charges can freely move—such as an aqueous solution, the uncanceled or net charges would not remain uniformly distributed. Rather, they repel each other and would therefore collect at the inner surface of the sphere.

The quantity c is therefore the hypothetical concentration of the net charge if it were uniformly distributed throughout the interior of the sphere.) The charge Q within the sphere of radius r is then $(4/3)\pi r^3 cF$. (The concentration) c can be in moles/cm^3, and here refers to the net concentration of uncompensated singly charged particles; F, which is in coulombs/mole, is necessary to convert from concentration to electrical charge.) The capacitance C of the sphere is $4\pi r^2 C'$, where $4\pi r^2$ is the surface area of a sphere and C′ is the capacitance per unit area. After substituting these values of Q and C Equation 1 and rearranging, we obtain the following expression for the electrical potential difference in the case of spherical capacitor :

$$\Delta E = \frac{rcF}{3C'} \qquad ...(2)$$

Equation 2 gives the electrical potential difference from the center of the sphere to just outside its surface. For a conductor, the internal uncompensated charges are found near the surface. In that case, ΔE actually occurs across the bounding surface (such as a membrane) surrounding the spherical body under consideration. Equation 2 indicates that the electrical potential difference developed is directly proportional to the average concentration of net charge enclosed, and inversely proportional to the capacitance per unit area of the sphere.

To apply Equation 2 to a specific situation, let us take a spherical cell with a radius of 30 μm and an electrical potential difference across the membrane (inside negative) of —100 mv, a value close to that occurring for many cells. If the membrane capacitance per unit area has a typical value of 1 μf/cm^2, to what net charge concentration in the cell does this potential drop correspond ? Since a farad is a coulomb/volt, one μf/cm ^{2} is 10^{-6} coulomb/v-cm^2 while F equals 96,487 coulombs/mole. Using Equation 2, c equals 3C′ ΔE/rF, hence the average concentration of net charge would be (3) (10^{-6} coulomb/v-cm^2) (– 0.1v)/[(30 × 10^{-4} cm) (96,487 coulombs/mole)], or –1.0 × 10^{-9} mole/cm^3, which is –1.0–10^{-6} mole/liter (—1 μM). The sign of the charge is negative, indicating more internal anions than cations.

The average concentration of the net (uncancealed) charge leading to a considerable electrical potential difference is rather small. It is instructive to compare the net charge concentration averaged over the volume of the cell (c in Equation 2 with the total concentration of anions and cations in a plant cell. Specifically, since the positive and negative ions in plant cells can each have a total concentration near 0.1 M, the above calculated excess of —1 μM is only about one extra negative charge per (0.1 M)/(10^{-6} M), or 10^5 anions. Expressed another way, the total charge of the cations inside such a cell equals or compensates that of the anions to within about one part in 100,000. When cations are taken up by a cell to any appreciable extent, either anions must accompany then and/or cations must be released from inside the cell. Otherwise, if marked departures from electrical neutrality were allowed to occur in some region, sizable electrical potential differences would build up.

Activity Coefficients of Ions

We will now turn our attention from the electrical term in the chemical potential to the activity one, specifically to the chemical activity itself. The activity of species j, aj, is the thermodynamically effective concentration. For charged particles in an aqueous solution this activity can differ appreciably from the actual concentration, cj—a point that has not always been adequately recognized in dealing with ions. A quantitative description of the dependence of the activity coefficient of ions on the concentration of the various species in the solution was developed by Debye and Huckel in the 1920s.

In a local region around a particular ion, the electrostatic forces describable by relations constrain the movement of other ions. As the concentration increases, the average distance between the ions becomes less, which facilitates ion-ion interactions. Equation electrical force=$Q_1Q_2/(4\pi\varepsilon_0 Dr^2)$] indicates, for example, that the electrostatic interaction between two charged particles varies inversely as the square of their separation, thus, electrical forces increase greatly as the ions get closer and closer together. When ions of opposite sign attract each other, the various other interactions of both ions are restricted, thus lowering their thermodynamically effective concentration or activity. A simple and approximate form of the Debye-Huckel equation appropriate for estimating the values of activity coefficients of ions in relatively dilute aqueous solutions at 25° is

$$\log\gamma_{+} = \frac{0.509\, z_{+}z_{-}\sqrt{\frac{1}{2}\sum_{j} cjzj^{2}}}{1+\sqrt{\frac{1}{2}\sum_{j} cjzj^{2}}} \qquad ...(3)$$

where z_+ is the charge number of the cation, z_- that of the anion, concentrations are expressed in molarities, and the summations are over all charged species. Thermodynamic measurements cannot give the activity coefficient of a single ion, because we cannot have a solution of one type of ion by itself in which to measure γ_+ or γ_-. In fact, activity coefficient of ions always occur as the products of those of cations and anions, hence $\gamma\pm$ in Equation 3 represents the mean activity coefficient of some cation-anion pair having charge numbers z_+ and z_-. Since z_- is negative, Equation 3 indicates that log $\gamma\pm$ is also negative, and $\gamma\pm$ is less than one. The activities of ions in dilute aqueous solutions are thus less than their concentrations, as we would indeed expect.

To estimate $\gamma\pm$ under conditions approximating those that might occur in a plant or animal cell, let us consider an aqueous solution containing 0.100 M of both monovalent cations and anions and 0.025 M of both divalent cations and anions. For this solution,

$$\frac{1}{2}\sum_{j} cjzj^{2}$$

(known as the ionic strength) is equal to $\frac{1}{2}[(0.100)(1)^2 + (0.100)(-1)^2$ $(0.025)(2)^2 + (0.025)(-2)^2]$, or 0.200 M. From Equation 3, log $\gamma\pm$ is then (0.509) (1) (–1) ($\sqrt{.0200}$)/(1+ $\sqrt{0.200}$), or –0.157 for the monovalent ions. This corresponds to a mean activity of only 0.70, a value considerably less than one.

The activity coefficient of a particular ionic species depends on all the ions in the solution, as is indicated by the ionic strength terms in Equation 3. Therefore, even when some particular ionic species is itself dilute, its activity coefficient can still be appreciably less than one because of the many electrostatic interactions with other ions. The departure from 1.00 for activity coefficients is even greater for divalent and trivalent ions than for monovalent ions, as the z_+z^- factor in Equation 3 indicates. Thus although activity coefficients of ions are often set equal to unity for convenience, this is obviously not always justified. A practical difficulty arising under most experimental situations is that cj is much easier to determine than aj, especially for compartments like the cytoplasm or the interior of a chloroplast. In those circumstances where aj has been replaced by cj, caution must be exercised in the interpretations or conclusions.

The activity coefficients for nonelectrolytes and water are generally much closer to unity than are those for ions, hence the assumption involved in replacing aj, by cj for such neutral species is not as severe as for the charged substances (e.g., $\gamma_{sucrose}$ is about 0.96 for 0.3 M sucrose). An all-inclusive theory for activity coefficients is rather complicated and beyond the scope of this text. Equation 3 is presented only to illustrate in a quantitative manner that activity coefficients of ions can be appreciably less than 1.00 under conditions that may occur in a plant cell.

Nernst Potential

Having considered the electrical and activity terms in some detail, we should now turn to the role of these quantities in the chemical potential of ions. Specifically, we will consider the deceptively simple and yet extremely important relationship between the electrical potential difference across a membrane and the accompanying distribution of ions across it at equilibrium. When ions of some species *j* are in equilibrium across a membrane, its chemical potential outside (o) is the same as inside (i), i.e., $\mu j^{\circ} = \mu j^{i}$. A difference in the Hydrostatic pressure term almost always makes a negligible contribution to the chemical potential differences of ions across membranes, therefore VjP can be omitted from μj in the present case. With this approximation and the definition of chemical potential the condition for equilibrium of ionic species *j* across the membrane is given by:

$$\mu j^* + RT \ln aj^{\circ} + zjFE^{\circ} = \mu j^* + RT \ln aj^{i} + zjFE^{i} \quad ...(4)$$

The term uj* in equation 4 is a constant referring to the same standard state of species *j* in the aqueous solutions on both sides of the membrane, and can be canceled from the equation.

Upon solving Equation 4 for the electrical potential difference, E^i — E°, across the membrane at *equilibrium*, we obtain the following important relationship :

$$E_{nj} = E^i - E^o = \frac{RT}{zjF} \ln \frac{aj^o}{aj^i}$$

$$= \frac{59.2}{zj} \log \frac{aj^o}{aj^i} \text{ mv at } 25° \qquad ...(5)$$

The electrical potential difference E_{Nj} in Equation 5 is called the *Nernst potential* of species *j*, i.e., $E^i–E^o = \Delta E = E_{Nj}$ in this case, where the subscript N stands for Nernst, who first derived this relation about 1900. We derived it by assuming equality of the chemical potentials of some charged species on two sides of a membrane, but the Nernst potential can be considered in a broader context (discussed below). For convenience of calculation, we can replace the natural logarithm (ln) by 2.303 log, where log is the common logarithm to the base 10. The quantity 2.303 RT/F equals 58.2 mv at 20°, 59.2 mv at 25° (cf. Eq. 5) and 60.2 mv at 30°.

Using such numerical values, we can express the nernst potential in the useful form presented in the second line of Equation 5, the Nernst equation, is an equilibrium statement showing how the internal and external activities of ionic species *j* are related to the electrical potential difference across the membrane. At equilibrium, a tenfold difference in activity of a monovalent ion across some membrane is energetically equivalent to a 59 mv difference in electrical potential (at 25°). Hence, a relatively small electrical potential drop can balance a large difference in activity across a membrane.

For some calculations, $\gamma j^o \gamma j^i$ is set equal to one (a less stringent assumption than setting both γj^o and γj^i equal to one). Under this condition, aj^o/aj^i in Equation 5 becomes the ratio of the concentrations, cj^o/cj^i ($aj = \gamma jcj$). Such a substitution may be justified when the ionic strengths on the two sides of the membrane are approximately the same, but it can lead to errors when the outside solution is much more dilute than the internal one, as would occur for *Chara* or *Nitella* in pond water. Throughout the rest of this book we will represent the actual electrical potential drop existing across a membrane, $E^i — E^r$, by E_M, where the subscript M refers to membrane. Hence, for both equilibrium, and non-equilibrium situations, we have $E^i—E^o=\Delta E = E_M$. When a particular ionic species *j* is in equilibrium across some membrane, E_M equals E_{Nj}, the Nernst potential for that species. However, regardless of the actual electrical potential difference existing across a membrane (E_M), a Nernst potential for an individual ionic species j can always be calculated from Equation 5 by using the ratio of the outside to the inside activity (aj^o/aj^i) or concentration (cj^o/cj^i), e.g., $E_{Nj} = (RT/zjF) \ln (cj^o/cj^i)$.

For many plant cells, the Nernst potential across the plasmalemma or the tonoplast calculated for potassium ions is close to the electrical potential differences measured across the membranes. If some ionic species cannot penetrate the membrane or is actively transported across it, E_{N_j} can markedly differ from E_M. In fact, the minimum amount of energy needed to transport ions across a membrane is proportional to the difference between E_{N_j} and E_M. The aqueous phases designated as inside and outside can have more than one membrane intervening between them. For instance, the vacuolar sap and an external solution—the regions often considered experimentally—have two membranes separating them, the plasmalemma and the tonoplast. The thermodynamic arguments remain the same in the case of multiple membranes, with E_M and E_{N_j} referring to the electrical potential differences between the two regions actually under consideration, regardless of how many membranes occur between them.

Example of E_{Nk}

Data obtained on potassium for the large internodal cells of the green freshwater alga *Chara australis* are convenient for illustrating the use of the Nernst equation (see Vorobiev). The potassium ion activity in the medium bathing the cells, a_K° (where the element symbol K as a subscript is the conventional shorthand notation for K^+) was 0.096 mM, while a_K^i in the vacuole was measured as 48 mM. Thus, a_k°/a_k^i was 0.096 mM)/48 mM), or 1/500. Using Equation 5 (with a factor of 58.2 since measurement was at 20°), the Nernst potential for potassium is (58.2 mv) log (1/500), or —157 mv. The measured electrical potential of the vacuole relative to the external solution. E_{M}, was –155 mv—very close to the calculated Nernst potential for potassium. Thus, potassium in this case may be in equilibrium between the external solution and the vacuole.

The potassium concentration in the vacuole of these *Chara* cells was 60 mM. The activity coefficient for potassium in the vacuole (γ_K^i) equals a_K^i/c_K^i ($a_j = \gamma_j c_j$, therefore γ_K^i was (48 mM)/(60 mM), or 0.80, while the potassium activity coefficient in the bathing solution (γ_K°) was about 0.96. For this example, where there are large differences in the internal and the external concentration, the ratio $\gamma_K^{\circ}/\gamma_K^i$ is (0.96)/(0.80), or 1.20, which differs appreciably from 1.00. If instead of activities concentrations had been used in Equation 5 (c_K°/c_K^i = 0.1/60 = 1/600), E_{Nk} would have been calculated as –162 mv, which is somewhat lower than the measured potential of –155 mv. Calculating from the concentration ratio, the suggestion that potassium is in equilibrium from the bathing solution to the vacuole could not have been made with much certainty, if at all.

Equilibrium does not require that the various forces acting on a substance are zero, rather, that they cancel each other.

In the above example for *Chara* the factors that tend to cause potassium to move are the differences both in its activity (or concentration) and the electrical potential across the membranes. The activity of potassium ions was much higher in the vacuole than in the external solution.

The activity term in the chemical potential therefore represents a driving force on potassium directed from the inside of the cell to the bathing solution. The electrical potential is lower inside the cell as hence the electrical driving force on the positively charged potassium ions tends to cause potassium entry into the cell. At equilibrium these two tendencies for movement in opposite directions are balanced and no net flux of potassium occurs. As indicated above, the electrical potential difference existing across a membrane when K^+ is in equilibrium is the Nernst potential of potassium, E_{NK}. Generally, $c_K{}^i$ and $a_K{}^i$ for both plant and animal cells are much higher than $c_K{}^o$ and $a_K{}^o$, and potassium is often close to equilibrium across the cellular membranes. From these observations, we can expect that the interiors of cells are usually at negative electrical potentials compared with the outside solutions, as is indeed the case. In general, the chemical potentials of ions are not equal in all regions of interest, hence passive movements toward lower μj occur—a topic we will turn to next.

Fluxes and Diffusion Potentials

Fluxes of many different solutes are continually occurring across biological membranes. Influxes take mineral nutrients into a growing cell while certain products of metabolism flow out of cells. The primary concern of this section will be the passive fluxes of ion toward lower chemical potential. Although the mathematical expressions on the next few pages are rather formidable, the underlying approach is really quite straightforward. First, we indicate that the flux of some species is directly proportional to the driving force causing the movement. Next, the driving force is expressed in terms of the relevant components of the chemical potential. We then examine the consequences of having electroneutrality when there are simultaneous passive fluxes of more than one type of ion. This leads us to an expression describing the electrical potential difference across a membrane in terms of the properties of the ions penetrating it. Before discussing the relation between fluxes and chemical potentials, we will briefly consider fluxes already mentioned or which may be familiar from other contexts.

In Fick's first law of diffusion, which says that the flux of (neutral) species j equals—$Dj\partial cj/\partial x$, where we can consider that the force is the negative gradient of the concentration. We are all familiar with Ohm's law in the form ΔE equals IR, where I is the current and R is the resistance across which the electrical potential drop is ΔE. The current per unit area A is simply the flux of charge, Jc, which equals —$(I/\rho)\partial E/\partial x$ where ρ is the electrical resistivity and the negative gradient of the electrical potential represents the driving force. We usually replace —$\partial E/\partial x$ by $\Delta E/\Delta x$, which leads to $I/A = (1/\rho)\Delta E/\Delta x$. In order to make this conform with the usual expression of Ohm's law (ΔE=IR), $\rho\Delta x/A$ must equal R, which indeed it does.

The gravitational force, —mjg, is simply the negative gradient of the potential energy in a gravitational field, i.e., —$\partial mjgh/\partial h = -mjg$, where the minus sign indicates that the force is directed toward decreasing altitudes, viz., toward the center of the earth. The gravitational force leads to the various forms of precipitation as well as to the percolation of water down through the soil. We have considered examples of fluxes depending on each of the variable terms of the chemical potential ($\mu j = \mu j^* + RT \ln aj + VjP + zjFE + mjgh$. In particular, the activity term (RT In aj) leads to Fick's first law, the pressure term (VjP) accounts for Darcy's law and Poiseuille's law, the electrical term (zjFE) yields Ohm's law, and the gravitational term (mjgh) is responsible for fluxes caused by gravity. In each case, flux is found by experiment to be directly proportional to the appropriate driving force.

We can actually generalize such relationships since nearly all transport phenomena can be represented by the statement : flux equals an appropriate force divided by some resistance (flux = force/resistance). Force is the negative gradient of a suitable potential, which we often conveniently take as the change in potential over some distance. But we have already shown that the chemical potential is an elegant way of summarizing all the factors which can contribute to the motion of a substance. It should not be surprising, therefore, that in general the flux of species *j* is proportional to the negative gradient of its chemical potential,—$\partial \mu j/\partial x$.

Flux and Mobility

Let us consider a charged substance, species j, which can cross a particular membrane. At equilibrium, the chemical potential of species j, μj, does not change with time or position—we are dealing with communicating regions-and there is no net flux of this solute across the membrane. When μj, changes with position but not with time, the situation is referred to as a *steady state*, a condition often used to approximate problem of biological interest. Actually, it is a matter of judgement whether μj is constant enough in time to warrant a description of the system as being in a steady state. (Similarly, constancy of μj for appropriate time and distance intervals is necessary before indicating that a system is in equilibrium.) Here we will let μj for appropriate time and distance intervals is necessary before indicating that a system is in equilibrium.) Here we will let μj change with both position and time over regions where species j is free to move; later we will restrict our development to the steady state case. When μj depends on position, a net passive movement or flux of species j, Jj will tend to occur toward the region where the chemical potential is lower.

The negative gradient of the chemical potential of species j, $-\partial\mu j/\partial x$, acts as the driving force for this Jj. The greater $-\partial\mu j/\partial x$ is, the larger will be the flux of species j in the x-direction. As a very useful approximation, we will assume that Jj is proportional to—$\partial\mu j/\partial x$, where the minus sign means that a net positive flux occurs in the direction of decreasing chemical potential. The magnitude of a flux across some plane is also proportional to the local concentration of that species, cj. That is, for a given driving force on species j, the amount of that substance which moves in proportional to how much of it is actually present—the more present, the greater the flux. Thus, for the one-dimensional case of crossing a plane perpendicular to the x-axis, Jj can be expressed as

$$
\begin{aligned}
Jj &= ujcj\left(-\frac{\partial uj}{\partial x}\right) \\
&= vjcj \qquad \text{...(6)}
\end{aligned}
$$

where uj is a coefficient called the mobility of species j. A mobility is generally the ratio of some velocity to the force which causes the motion—we will return to a consideration of uj shortly.

As we have already indicated, the top line of Equation 6 is a representative example from the large class of expressions relating various flows to their causative forces. In this particular case, Jj is the rate of flow of moles of species *j* across unit area of a plane, and can be expressed in moles/cm^2-sec. Such a molar flux of species j divided by its local concentration, cj, gives the mean velocity, vj, with which this species moves across the plane—when cj is in moles/cm^3, Jj/cj can have units of (moles/cm^2-sec)/(moles/cm^3) or cm/sec. Perhaps this important point can be better appreciated by considering it in the following way.

Let us suppose that the average velocity of species j moving perpendicularly toward area A of the plane of interest is vj, and we will consider a volume element of cross-sectional area A which extends back from the membrane for a distance vjdt. In a time dt, all molecules in the volume element vjdtA will cross area A, which mean that the number of moles of species j crossing in this interval is (vjdtA) × (cj), where cj is the number of moles/unit volume. Hence, the molar flux, Jj (which is simply the number of moles crossing unit area per unit time), would be vjdt Acj/(Adt), or vjcj. In other words, the mean velocity of species j moving across the plane (vj) times the number of those molecules per unit volume which can move (cj) equals the flux of that species (Jj), as is given by the bottom line of Equation 6.

The upper line of Equation 6 indicates that this average velocity, Jj/cj, equals the mobility of species j, uj, times $—\partial\mu j/\partial x$, the latter being the force on j that causes it to move. Thus, mobility is simply the proportionality factor between the mean velocity of motion (vj=Jjcj) and the causative force $(-\partial\mu j/dx)$. The greater of mobility of some species, the larger its velocity in response to a given force. The particular form of the chemical potential of species j, μj, to be substituted into Equation 6 depends on the specific application that we have in mind. For charged particles moving across biological membranes, the appropriate μj is $\mu j^* + RT \ln aj + zjFE$. (As we mentioned before, the VjP term makes only a relatively small contribution to the $\Delta\mu j$ of an ion, therefore it is not included.

For treating the one-dimensional case described by Equation 6, μj must be differentiated with respect to x, $\partial\mu j/\partial x$, which leads to $RT\partial \ln aj/\partial x + zjF\partial E/\partial x$ when T is constant. The quantity $\partial \ln aj/\partial x$ equals $(1/aj)\partial aj/\partial x$, which is $(1/\gamma jcj/\partial x(aj= \gamma jcj)$. Using the above form of μj appropriate for charged solutes and this expansion of the derivative, $\partial \ln aj/\partial x$, the net flux of species j appearing in Equation 6 can be written

$$Jj = -\frac{u_j RT}{\gamma_j}\frac{\partial \gamma_j c_j}{\partial x} - u_j c_j z_j F\frac{\partial E}{\partial x} \qquad ...(7)$$

Equation 7 becomes simply $- ujRT(\partial cj/\partial x)$, i.e., it is proportional to the j either across a membrane or in a solution in terms of two components of the driving force : the gradients in activity and electrical potential.

Before proceeding, we should examine a little more closely the first term on the right side of Equation 7. When γj varies across the membrane, $\partial\gamma_j/\pi x$ can be considered to represent a driving force on species j. In keeping the with common practice, we will ignore this possible force, i.e., we will assume that γj can be treated as a constant — in any case

$\partial\gamma j/\partial x$ would be extremely difficult to measure. For constant γj, the first term on the right side of Equation 7 becomes simply—ujRT(∂cj/∂xj, i.e., it is proportional to the concentration gradient. In the absence of electrical potential gradients ($\partial E/\partial x = 0$), and for neutral solutes (zj = 0), Equation 7 indicates that Jj equals—ujRT(∂cj/∂x). But this is just the type of flux described by Fick's first law (Jj=-Dj∂cj/∂x), with ujRT taking the place of the diffusion coefficient, Dj. In other words, we can replace ujRT in Equation 7 by Dj. Since Dj equals ujRT, diffusion coefficients must depend on temperature. Moreover, uj is generally inversely proportional to the viscosity, which decreases as T increases, thus the dependence of Dj on T can be pronounced. Consequently, the temperature should be specified when the value of diffusion coefficient is given. Let us now return to a consideration of fluxes.

In the absence of electrical effects and for constant γj, the net flux of species j given by Equation 7 equals — Dj∂cj/∂x when ujRT is replaced by Dj. Thus, Fick's first law (Jj = – (Jj = Dj∂cj/∂x, is a special case of our general flux relation (Equation 6), where we ignored first the pressure and gravitational effects in order to obtain Equation 7 and then the electrical effects. But this eventual reduction of Equation 6 to Fick's first law is as we should expect, since in fact the only driving force considered when we presented Fick's first law was the concentration gradient. Such agreement between our present thermodynamic approach and the seemingly more empirical Fick's first law is quite important. It serves to justify the logarithmic term for activity in the chemical potential (μj = μj* + RT ln aj + VjP + zjFE + mjgh. In other words, if the activity of species *j* appeared in the above Equation in a form other than ln aj. Equation 7 would not reduce to Fick's first law under the appropriate conditions.

Since Fick's first law has been amply demonstrated experimentally, such a disagreement between theory and practice would necessitate some modification in the expression used to define chemical potential. The terms in the chemical potential can be justified or "derived" by different methods. The forms of some terms in μj can be readily appreciated since they follow from familiar definitions of work, e.g., the gravitational term and the electrical term. The above comparison with Fick's first law indicates that the ln aj form is the appropriate way of handling the activity term. Some of these derivations incorporate conclusions from empirical observations.

Moreover, the fact that the chemical potential can be expressed as a series of terms which can be added together agrees with experiment. Thus a thermodynamic expression for the chemical potential can (1) summarize the results of previous observations (2) stand the test of experiments, and (3) lead to new and useful predictions. In contrast to the case for a neutral solute, the flux of an ion also depends on an electrical driving force, represented in Equation 7 by $-\partial E/\partial x$. A charged solute spontaneously tends to move in an electrical potential gradient — a cation moving in the direction of lower or decreasing electrical potential.

Of course, the concentration gradient also affects charged particles. If a certain type of ion were present in some region but absent in an adjacent one, the ions would tend to diffuse into the latter region. A such charged particles diffuse toward regions of lower concentration, an electrical potential difference is created. This electrical potential difference is referred to as a "diffusion potential," which we will discuss in detail below. The interrelationship between concentration and electrical effects is extremely important in biology. Most membrane potentials can be treated as diffusion potentials resulting from different rates of movement of the various ions across a membrane.

Diffusion Potential in a Solution

We will now use Equation 7 to derive the electrical potential difference created by ions diffusing down a concentration gradient in a solution containing one type of cation and its accompanying anion. This case is relatively simple compared with the more biological one to follow, thus it may more clearly illustrate the relationship between concentration gradients and the accompanying electrical potential differences. We will assume that the cations and anions are initially placed at one side of the solution. In time, they will *diffuse* across the solution toward regions of lower concentration.

In general, one ionic species will have a higher mobility, u_j, than the other. The more mobile ions will tend to diffuse faster than their oppositely charged partners, resulting in a microscopic charge separation. This slight charge separation sets up an electrical potential gradient leading to a *diffusion potential*. Using certain simplifying assumptions, we will calculate the magnitude of the electrical potential difference so created. For convenience of analysis, let us consider a solution containing only a monovalent cation (+) and its monovalent anion (–). We will assume that their activity coefficients are constant.

As the previous calculation on electrical effects indicated, solutions are essentially neutral in regions which are large compared with atomic dimensions. Thus, c_+ equals c_-, and the concentration of either species can be designated c. Furthermore, no charge imbalance will develop in time, which means that the flux of the cation across some plane in the solution, J_+, equals that of the anion across the same plane, J_-. (A very small charge imbalance does develop, which sets up the electrical potential gradient, but this uncompensated flux is transitory and in any case negligible compared with J_+ or J_-.) Both fluxes, J_+ and J_-, can be expressed by Equation 7 and then equated to each other, with gives

$$-u_+RT=\frac{\partial c}{\partial x}-u_+cF\frac{\partial E}{\partial x}=-u_-RT\frac{\partial c}{\partial x}+u_-cF\frac{\partial E}{\partial x} \quad ...(8)$$

where the plus sign on the right side occurs because the monovalent anion carries a negative charge ($z^- = -1$). Rearrangement of Equation 8 yields the following expression for the electrical potential gradient:

$$\frac{\partial E}{\partial x}=\frac{u_- - u_+}{u_+ + u_-}\frac{RT}{Fc}\frac{\partial c}{\partial x} \quad ...(9)$$

Equation 9 indicates that a nonzero $\partial E/\partial x$ occurs when the mobility the cation differs from that of the anion and there is a concentration gradient. If u^- is greater than u_+, the anions move (diffuse) faster than the cations toward regions of lower concentration. As some individual anion moves ahead of its "partner" cation, an electric field is set up in such a direction as to speed up the cation and slow down the anion until they both move at the same speed, thus preserving electrical neutrality.

In order to obtain the difference in electrical potential produced by diffusion between planes of differing concentration, we must integrate Equation 9. At this point we will restrict our consideration to the steady state case where neither E nor c changes with time. In going along the x axis from region I to region II, the change in the electrical potential term in Equation 9 is $\int_I^{II}(\partial E/\partial x)dx$, which becomes $\int_I^{II}dE$ for the steady state condition and so equals $E^{II} - E^{I}$. The integral of the concentration term is $\int_I^{II}(1/c)(\partial/\partial x)dx$, which becomes $\int_I^{II}dc/c$, or ln (c^{II}/c^{I}). Using these two relations, integration of Equation 9 leads to

$$E^{II}-E^{I}=\frac{u_- - u_+}{u_+ + u_-}\frac{RT}{F}\ln\frac{c^{II}}{c^{I}}$$

$$=59.2\frac{u_- - u_+}{u_+ + u_-}\log\frac{c^{II}}{c^{I}}\text{ mv at }25° \quad ...(9)$$

where in the second line ln has been replaced by 2.303 log, and the value 59.2 mv at 25° has been substituted for 2,303 RT/F. In the general case, the actions have mobilities different from the cations. As the ions diffuse to regions of lower chemical potential in the solution, an electrical potential difference—given by Equation 9, and called a diffusion potential

—is set up across the section where the concentration changes from c^I to c^{II}.

An example of a diffusion potential describable by Equation 9 occurs at the open end of the special micropipettes used for measuring electrical potential differences across membranes. (The "pulling," filling, and using of micropipettes—commonly referred to as microelectrodes or microsalt-bridges. The fine tip of the micropipette is open and provides an electrically conducting pathway into the cell or tissue. Ions diffusing from this fine tip give rise to a diffusion potential between the interior of the micropipette and the aqueous compartment into which the tip is inserted.

To estimate the magnitude of this potential, we will assume that there is a concentration ratio of 30 across the micropipette tip, and then calculate the diffusion potentials for typical electrolytes, NaCl and KCl. The chloride mobility, uc_l, is about 1.52 times u_{Na} and so $u^- - u_+)/(u_+ + u^-)$ is (0.52/(2.52), or 0.21 for NaCL. When the concentration ratio is 30, the diffusion potential calculated by Equation 9 is (59.2 mv)(0.21)log ($^1/_{30}$), or — 18mv. For KCl as the electrolyte, uc_l is 1.04 u_K, and thus $(u_- - u_+)/(u_+ + u_-)$ is (0.04)/(2.04), or 002. This leads to a diffusion potential of (59.2)(0.02) log ($^1/_{30}$) or only –2 mv in going from the interior of the micropipette, where the concentration of KCl is, say 3M, to a region where it is 0.1 M. Thus, from the point of view of minimizing the diffusion potential across the fine tip, KCl is a much more suitable electrolyte for micropipettes than is NaCL. In fact, KCl is employed in nearly all micropipettes used for measuring membrane potentials (generally c_K inside cells is also higher than c_{Na}, and it does not vary as much from cell to cell as does c_{Na}). In any case, the closer u is to u_+, the smaller will be the diffusion potential predicted by Equation 9 for a given concentration ratio.

Membrane Fluxes

As is the case for diffusion potentials in open solution, membrane potentials also depend on the different mobilities of the various ions and on their concentration gradients. In this case, however, the "solution" in which the diffusion takes place toward regions of lower chemical potential is actually the membrane itself. In this regard, we should mention were that a membrane is often the main barrier, or rate-limiting step for the diffusion of molecules into and out of cells or organelles. We would therefore except it to be the phase across which the diffusion potential is expressed.

Under biological conditions a number of different types of ions are present, hence the situation is more complex than for the single cation-anion pair analyzed above. Furthermore, the quantities of interests such as $\partial E/\partial x$ and γ_j are those within the membrane, hence they are not accessible to experimental measurement. We will consider the ways of dealing with these difficulties in this section. In order to calculate membrane diffusion potentials, we must make certain assumptions before an integration of fluxes or gradients can be made. As a start, we will assume that the electrical potential (E) varies linearly with distance across the membrane. This means that $\partial E/\partial x$ is a constant equal to $E_M/\Delta x$, where E_M is the electrical potential difference across the membrane —i.e., $E^i - E^o = \Delta E = E_M$, and Δx is the membrane thickness. This assumption of a constant electric field across the membrane—the electric field equals —$\partial E/\partial x$ in our one-dimensional case—was originally suggested by D.E. Goldman in 1943. It appreciably simplifies the integration conditions leading to an expression describing the electrical potential difference across biological membranes.

As another very useful approximation, we will assume that the activity coefficient of species j, γ_j, is constant across the membrane (a less severe restriction than setting it equal to one) A partition coefficient is needed to describe concentrations within a membrane since the solvent properties of a membrane are different from those of the aqueous solution on either side of it, where the concentrations are actually determined. Thus, c_j in Equation 7 should be replaced by K_jc_j, where K_j is the partition coefficient for species j. Incorporating these various simplifications and conditions, and after transferring the electrical term to the left side, we can rewrite Equation 7 as follows:

$$u_jK_jc_jF\frac{E_M}{\Delta x}+J_j=-u_jRTK_j\frac{\partial c_j}{\partial x} \qquad ...(10)$$

In order to cast Equation 10 into a form that can be integrated, it is advantageous to pull all terms containing the concentration on the same side of the equation. Furthermore, since the two most convenient variables are c_j and x, their differentials should appear on opposite sides of the equation. Being guided by these two objectives, and after first multiplying each term by $dx/(u_jRT)$, we can rearrange Equation 10 to give the following expression:

$$\frac{z_jFE_M}{RT\Delta x}dx=-\frac{K_jdc_j}{K_jc_j+\dfrac{J_j\Delta x}{u_jz_jFE_M}} \qquad ...(11)$$

where $(\partial c_j/\partial x)\,dx$ has been replaced by dc_j in anticipation of the restriction to steady-state conditions.

When the term $J_j\Delta x/(u_j z_j EF_M)$ is constant, Equation 11 can be readily integrated from one side of the membrane to the other. The factors Δx, z_j, F, and E_M are all fixed, and only J_j and u_j need be further considered here. When the flux of species *j* does not change with time or position, then J_j across any plane parallel to and within the membrane is the same, and species *j* is not accumulating or being depleted in any of the regions of interest. Consequently, $\partial c_j/\partial t$ is zero, while of course $\partial c_j/\partial x$ is nonzero, which is the steady state condition.

Our restriction to a steady state therefore means that J_j is constant and $(\partial c_j/\partial x)$ equals dc_j, a relation which we have already incorporated into Equation 11. For convenience, the mobility (uj) of each species is also assumed to be constant within the membrane. With these restrictions, the quantity $J_j\Delta x/c_j z_j FE_{M)\ \text{becomes}}$ a constant, and we can integrate Equation 11 from one side of the membrane (the outside, o) to the other (the inside *i*).

We know $\int dx_i\,(x+b)$ equals $\ln(x+b)$: Hence,

$$-\int_{K_j c_j^{o}}^{K_j c_j^{i}} K_j dc_j/(K_j c_j + b)$$

is

$$-\ln(K_j c_j + b)\Big|_{K_j c_j^{o}}^{K_j c_j^{i}}$$

or $-\ln(K_j c_j^{i} + b)/(K_j c_j^{o} + b)$, which equal $\ln(K_j c_j^{o} + b)/(K_j c_j^{i} + b)$. Since

$$\int_{z^{o}}^{z^{i}} dx \text{ equals } \Delta x,$$

integration of Equation 11 gives

$$\frac{z_j FE_M}{RT} = \ln \frac{\left(K_j c_j + \dfrac{J_j \Delta x}{u_j z_j FE_M}\right)}{\left(K_j c_j^{i} + \dfrac{J_j \Delta x}{u_j z_j FE_M}\right)} \qquad ...(12)$$

After taking exponentials of both sides of Equation 12 in order to put it into a more convenient form, and then multiply by $K_j c_j^{i} + J_j\Delta x/(u_j z_j FE_M)$, Equation 12 becomes

$$K_j c_j^i \, e^{z_j F E_M/RT} + \frac{J_j \Delta x}{u_j z_j F E_M} e^{z_j F E_M/RT}$$

$$= K_j c_j^{\circ} + \frac{J_j \Delta x}{u_j z_j F E_M} \quad ...(13)$$

A quantity of considerable interest in Equation 13 is J_j, the net flux of species *j*. This equation can be readily solved for J_j, giving

$$J_j = J_j^{in} - J_j^{out} = \left(\frac{K_j u_j z_j F E_M}{\Delta x}\right)\left(\frac{1}{e^{z_j F E_M/RT} - 1}\right)$$

$$(c_j^{\circ} - c_j^i \, e^{z_j F E_M/RT}) \quad ...(14)$$

where J_j^{in} is the influx of species j, J_j^{out} is its efflux, and their difference is the net flux. (The net flux J_j can represent either a net influx or a net efflux, depending on which of the unidirectional components, J_j^{in} or J_j^{out}, is larger.) Although the mathematical manipulations necessary to go from Equation 10 to Equation 14 are lengthy and cumbersome, the resulting expression is of extreme importance for understanding both membrane potentials and the passive fluxes of ions. Moreover, throughout this text all the steps involved in a particular derivation have been shown in order to avoid statements such as "it can easily be shown that such and such follows from so and so"—expressions which can be most frustrating and often are untrue.

Equation 14 will be used to derive the Goldman equation—describing the diffusion potential across membranes—and will be invoked later in the derivation of the Ussing-Teorell equation, a relation, obeyed by certain passive fluxes. Equation 14 shows how the passive flux of some charged species *j* depends on its internal and external concentrations and on the electrical potential difference across the membrane. For most cell membranes, E_M is negative, i.e., the inside of the cell is at a lower electrical potential than is the outside. For a cation (zj a positive integer) and a negative E_M, the terms in the first two parentheses on the right side on Equation 14 are both negative, hence their product is positive.

On the other hand, for an anion (z_j a negative integer) and negative E_M, both parentheses are positive, and their product is still positive. (For $E_M > 0$, the product of the first two parentheses is also positive for both anions and cations.) Thus, the sign of J_j depends on the value of c_j° relative to that of $c_j^i \, e^{z_j F E_M/RT}$. When c_j° is greater than $c_j^i \, e^{z_j F E_M/RT}$, the expression in the last parenthesis of Equation 14 is positive, and a net inward flux of species j occurs ($J_j > 0$). Such a condition may be contrasted with Equation [$J_j = P_j(c_j^{\circ} - c_j^{i})$], where the net flux in the absence of electrical effects was inward when c_j° was simply larger than c_j^{i}, as would adequately describe the situation for neutral solutes. However, knowledge of the concentration difference alone is not sufficient to predict the magnitude or even the direction of the flux of ions; we must also consider the electrical potential difference between the two regions.

Membrane Diffusion Potential—Goldman Equation

Passive fluxes of ions, which can be described bp Equation 14 and are caused by gradients in the chemical potentials of the various species, lead to an electrical potential difference (diffusion potential) across a membrane. We can determine the magnitude of this electrical potential difference by considering the contributions from all ionic fluxes across the membrane and the condition of electroneutrality. Certain assumptions are needed, however, to keep the equations manageable. Under usual biological situations not all anions and cations can easily move through the membranes.

Many divalent cations do not readily enter or leave plant cells passively, which means that their mobility in the membranes is small. Such ions usually do not make a large enough contribution to the fluxes into or out of plant cells to influence markedly the diffusion potentials across the membranes. Thus, we will omit them in the present analysis, which is nevertheless rather complicated. For many plant cells, the total ionic flux consists mainly of movements of K^+, Na^+, and Cl^- . These three ions generally have fairly high concentrations in and near plant cells, and we would therefore expect them to make substantial contributions to the total ionic flux.

More specifically, a flux of species j depends on the product of its concentration and its mobility ($J_j = -u_j c_j \partial\mu_j/\partial x$, Equation 6, thus, ions having relatively high local concentrations or moving in the membrane fairly easily (high u_j) will tend to be the major contributors to the total ionic flux. In some cases there may be a sizable flux of H^+ or OH^- (which can have high u_js) as well as other ions, the restriction here to three ions being partially in the interest of algebraic simplicity. However, the real justification for considering only K^+, Na^+, and Cl^- is that the diffusion potentials calculated by using the passive fluxes of these three ions across biological membranes are often in good agreement with the measured electrical potential differences. Our previous electrical calculations indicated that solutions are essentially neutral, a condition which we will again invoke here. In other words, because of the large effects resulting from small amounts of uncanceled charge, the net charge actually needed to cause the electrical potential difference across a membrane (E_M) is negligible compared with the existing concentrations of ions. Furthermore, the steady-state fluxes of the ions across the membrane will not change this condition of electrical neutrality, which means that no net charge is transported by the algebraic sum of the various charge movements across the membrane, i.e.,

$$\sum_j z_j J_j = 0.$$

When the bulk of the ionic flux consists of K^+, Na^+, and Cl^- movements this important condition of electroneutrality can be described by equating the cationic fluxes ($J_K + J_{Na}$) to the anionic one (J_C), which leads to the following relation (as mentioned above, the various ions are indicated by using only their element symbols as subscripts)

$$J_K + J_{Na} - J_{Cl} = 0 \qquad \text{...(15)}$$

Equation 15 describes the net ionic fluxes which lead to the electrical potential difference across the membrane. After substituting the expressions for the various J_js into Equation 15, we will solve the resulting equation for the diffusion potential across the membrane, E_M.

To obtain a useful expression for E_M in terms of measurable parameters, it is convenient to introduce a permeability coefficient for species *j*, P_j. In such a permeability coefficient was defined as $D_jK_j/\Delta x$, where D_j is the diffusion coefficient of species *j*, K_j is its partition coefficient, and Δx is the membrane thickness. Upon comparing Equation 7 with ($J_j = -D_j\partial c_j/\partial x$), we can see that u_jRT takes the place of the diffusion coefficient of species j, D_j. The quantity $K_ju_jRT/\Delta x$ can thus be equated to $K_jD_j/\Delta x$, which is the permeability coefficient (P_j) of species j in a given membrane, the unknown thickness of the membrane, and the partition coefficient for the solute can be replaced by one parameter, which describes the permeability properties of that solute crossing the particular membrane under consideration. With all the preliminaries out of the way, let us now determine the expression for the diffusion potential across a membrane for the case where most of the ionic flux is due to K^+, Na^+, and Cl^- movements. Using the permeability coefficients of the three ions and substituting in the net flux of each species as defined by Equation 14, Equation 15 becomes

$$P_K\left(\frac{1}{e^{FE_M/RT}-1}\right)\left(C_{K^\circ} - C_K e^{FE_M/RT}\right) + P_{Na}\left(\frac{1}{e^{FE_M/RT}-1}\right)\times$$

$$\left(C_{Na^{o}} - C_{N_a^{i}} e^{FE_M/RT}\right) + P_{Cl}\left(\frac{1}{e^{-FE_M/RT} - 1}\right)\left(C_{C_i^{o}} - C_{C_i^{i}} e^{-FE_M/RT}\right)$$
$$= 0 \qquad ...(16)$$

where z_K and z_{Na} have been replaced by 1 z_{Cl} by –1, and FE_M/RT has been canceled from each of the terms for the three fluxes. To simplify this rather unwieldy expression, the quantity 1/(e $^{FEM/RT}$ –1) can be canceled from each of the three terms represented in Equation 16 1/(e$^{-FEM/RT}$ –1) appearing in the last term is the same—as $e^{FEM/RT}/e^{FEM/RT}$ –1). Equation 16 then assumes a; much more manageable form:

$$P_{K}C_{K^{o}} - P_{K}C_{K^{i}} e^{FE_M/RT} + P_{Na}C_{N_a^{i}} e^{FE_M/RT}$$
$$-P_{Cl}C_{C_i^{o}} e^{FE_M/RT} + P_{Cl}C_{C_i^{i}} = 0 \qquad ...(17)$$

After solving Equation 17 for $e^{FEM/RT}$ and taking logarithms, we obtain the following expression for the electrical potential difference across the membrane :

$$E_M = \frac{RT}{F} \ln \frac{P_{K}C_{K^{o}} + P_{Na}C_{Na^{o}} + P_{Cl}C_{C_i^{i}}}{(P_{K}C_{K^{i}} + P_{Na}C_{Na^{i}} + P_{Cl}C_{Ci^{o}})} \qquad ...(18)$$

Equation 18 is generally known as the Goldman, or constant field equation. The electric field equals —∂E/∂x, which Goldman in 1943 set equal to a constant (here—$E_M/\Delta x$) in order to facilitate the integration across a membrane. It turns out that the assumption of a constant electric fields in the membrane is actually not essential for obtaining Equation 18, we could invoke Gauss's law and perform a more difficult integration. In 1949, Hodgkin and Katz applied the general equation derived by Goldman to the specific case of K^+, Na^+, and Cl^- diffusing across a membrane, and Equation 18 is often referred to as the Goldman-Hodgkin-Katz equation.

Equation 18 gives the diffusion potential existing across a membrane. We have derived it by assuming independent passive movements of K^+, Na^+ and Cl^- across a membrane in which ∂E/∂x, γ_j, J_j, and u_j are all constant. We used the negative gradient of its chemical potential as the driving force for the net flux of each ion. Thus, Equation 18 gives the electrical potential difference arising from the different tendencies of K^+ Na^+, and Cl^- to *diffuse* across a membrane to regions of lower chemical potential. When other ions cross some particular membrane in appreciable amounts they will also make a contribution to its membrane potential. However, the inclusion of divalent and trivalent ions in the derivation of an expression for E_M complicates the algebra considerably (e.g., if Ca^{++}

is also considered Equation 17 has fourteen terms on the left side instead of six, and the equation becomes a quadratic in powers of $e^{EFM/RT}$). The fluxes of such ions may be rather small, in which case Equation 18 may be adequate for describing the membrane potential.

Application of the Goldman Equation

In certain cases, the terms appearing in Equation 18, viz., the permeabilities and the internal and external concentrations of K^+, Na^+, and Cl^-, have all been measured. The validity of the Goldman equation can then be checked. In other words, we can compare the predicted diffusion potential with the actual electrical potential difference experimentally measured across the membrane. As a specific example, we will use the Goldman equation to evaluate the membrane potential across the plasmalemma of *Nitella translucens*.

The concentrations of K^+, Na^+, and Cl^- in the external bathing solution and in the cytoplasm of Nitella translucens. The ratio of permeability of Na^+ to that of K^+, P_{Na}/P_{K}, is about 0.18 for *Nitella translucens*. The Plasmalemma of *Nitella* is much less permeable to Cl^- than it is to K^+, probably only 0.1% to 1% as much. Thus, for purposes of calculation, we will let P_{Cl}/P_K be 0.0003. In terms of permeabilities relative to that for potassium, P_K for *Nitella* is then 1.00 P_{Na} is 0.18 and P_{Cl} is 0.003. Using these relative permeability coefficient and the argument of the logarithm appearing in Equation 18, $(P_{KCK}{}^{o} + P_{NoCNo}{}^{o} + P_{ClCCl}{}^{i})/(P_{KCK}{}^{i} + P_{NaCNa}{}^{I} + P_{ClCCl}{}^{o})$, has the numerical value of [(1.0)(0.1 mM) + (0.18)(1.0 mM) + (0.003) (65m M)] divided by [(1.0)(119 mM) + (0.18)(14 mM) + (0.003)(1.3 mM), which is (0.475)/(121.5), or 0.00391. Since in is 2.303 log and 2.303 RT/F equals 58.2 may at 20°, the Goldman equation predicts a membrane potential, EM, of (58.2 mv) log (0.00391), or –140 mv. Thus, we expect the cytoplasm to be electrically negative with respect to the external bathing solution, as is indeed the case.

In fact, the measured value of the electrical potential difference across the plasmalemma of *Nitella translucens* is – 138 mv at 20°. This excellent agreement between the observed electrical potential difference and that calculated from the Goldman equation supports the contention that the membrane potential is in fact a diffusion potential. This point can be fairly easily checked by varying the external concentration of K^+, Na^+, and/or Cl^- and then seeing whether the membrane potential changes in accordance with Equation 18.

The, different ionic concentration on the two sides of a membrane help set up the passive ionic fluxes creating the diffusion potential, but the actual magnitude of the contribution of a particular ionic species to

E_M also depends on the ease with which that type of ion can cross the membrane. Based on the relative permeabilities and concentration, the major contribution to the electrical potential difference across the plasmalemma of Nitella comes from the K flux, with Na and Cl playing secondary roles. If the Cl terms are omitted from Equation 18 (i.e., if P_{Cl} is set equal to zero), the calculated membrane potential is – 153 mv, compared with –140 mv when Cl is included. This rather small difference between the two potentials is a reflection of the relatively low permeability coefficient for chloride crossing the plasmalemma of Nitella, hence the Cl flux has less effect on E_M than does the K flux. Changes in the amount of Ca^{++} or H^+ in the external medium cause some deviations from the predictions of Equation 18 for the electrical potential difference across the plasmalemma of *Nitella.* Thus, the diffusion potential is influenced by the properties of the particular membrane being considered, and ions other than K^+, Na^+, and Cl^- may have to be included in specific cases.

To allow for the influence of the flux of H^+ on E_M, for example, we could include $P_{HCH}{}^o$ in the numerator of the logarithm in the Goldman equation (Equation 18) and $P_{HCH}{}^i$ in the denominator (note that a movement of H^+ in one direction has the same effect on E_M as movement of OH^- in the other). Nevertheless, restriction to the three ions indicated has proved to be adequate for treating the diffusion potential across many membranes, and the Goldman, or constant field expression in the form of Equation 18 has found widespread application.

Donnan Potential

Another type of electrical potential difference encountered in biological systems is associated with immobile or fixed charges in some region adjacent to an aqueous phase containing small mobile ions, and is referred to as a *Donnan potential.* When a plant cell is placed in a kcl solution, for example, a Donnan potential arises between the interior of the cell wall and the bulk of the bathing fluid. The electrical potential difference arising from electrostatic inter –actions at such a solid-liquid interface can be regarded as a special type of diffusion potential, as we will show shortly. Pectin and other macromolecules in the cell wall have a large number of immobile carboxyl groups (—COOH) from which hydrogen ions dissociate. This gives the cell wall a net negative charge, as indicated. Cations such as Ca^{++} are electrostatically attracted to the negatively charged cell wall, and the overall effect is an exchange of H^+ for Ca^{++} and other cations. The region containing the immobile charges—such as carboxyl groups in the case of the cell wall—is generally referred to as the *Donnan phase.*

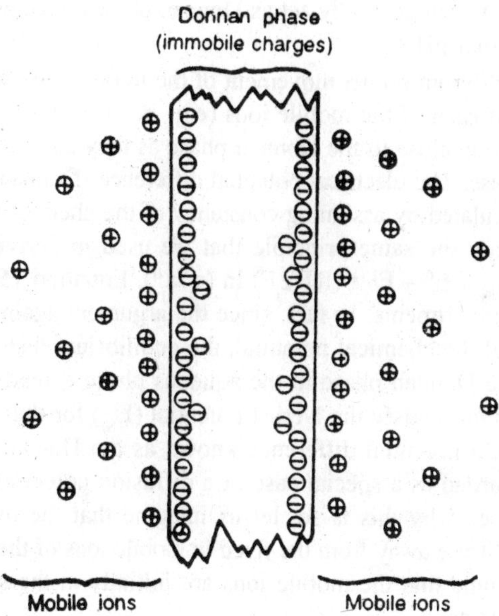

Fig. 4.3. Spatial distribution of positively charged mobile ions ⊕ occurring on either side of a Donnan phase in which are embedded immobile negative charges (Θ).

At equilibrium, a distribution of oppositely charged ions electrostatically attracted to these immobile charges occurs between the Donnan phase and the adjacent aqueous one . This sets up an electrical potential gradient and a Donnan potential is thus created between the center of the Donnan phase and the bulk of the solution next to it. The sign of the electrical potential in the Donnan phase relative to the surrounding electrolyte solutions is the same as the sign of the charge of the immobile ions. For example, because of the presence of dissociated carboxyl groups, the electrical potential in the cell wall is negative with respect to an external solution. Membranes also generally act as charged Donnan phases. In addition, Donnan phases occur in the cytoplasm where the immobile charges are due to proteins and other large polymers (RNA, DNA) that have many carboxyl and phosphate groups from which protons can dissociate, leaving the macromolecules with a net negative charge. Such polymers are fixed in the sense that they are unable to diffuse across either the plasmalemma or the tonoplast. When the immobile or fixed ions are in some region, mobile ions of opposite sign can occur in adjacent layers on either side. For example, layers containing cations are often formed in the aqueous solutions on each side of biological membranes, which generally act as Donnan phases having a net negative charge at usual pH's.

At equilibrium no net movement of the ions occurs, so the chemical potentials of each of the mobile ions (e.g., K^+, Na^+, Cl^- , Ca^{++}) have the same values up close to the Donnan phase as they do in the surrounding aqueous phase. The electrical potential difference (Donnan potential) can then be calculated by assuming constancy of the chemical potential. But this is exactly the same principle that we used in deriving the Nernst potential, $E_{Nj} + E^{II} - E^{I} = (RT/z_jF)$ In (a_j^I/a_j^{II}). Equation 15, between two aqueous compartments. In fact, since the argument again hinges on the constancy of the chemical potential, the equilibrium distribution of any ion from the Donnan phase to the aqueous phase extending away from the barrier must satisfy the Nernst potential (E_{Nj}) for that particular ion. The electrical potential difference known as the Donnan potential can also be regarded as a special case of a diffusion potential, as suggested above. To see why this is so, let us imagine that the mobile ions are tending to diffuse away from the fixed immobile ions of the opposite sign. We can assume that the mobile ions are initially in the same region as the immobile ones.

In time, the mobile ions would tend to diffuse away. This tendency, based on thermal agitation, causes a slight charge separation and thus sets up an electrical potential difference between the Donnan phase and the bulk of the adjacent solution. For the case of a single species of mobile cations and the anions fixed in the membrane (both assumed to be monovalent here), the diffusion potential across that part of the aqueous phase next to the membrane can be described by Equation 9, $E^{II} - E^{I} = [(u_- - u_+)/(u_+ + u_-)]$ (RT/F) In (c^{II}/c^{I}), which we derived for diffusion toward regions of lower chemical potential in a solution. Fixed anions have zero mobility ($u_- = 0$), hence $(u_- - u_+)/ (u_+ + u_-)$ here is $- u_+/ u_+$, or -1. Equation 9 then becomes $E^{II} - E^{I}$ equals $-$ (RT/F) ln (c^{II}/c^{I}), which is simply the Nernst potential (Equation 5) for monovalent cation [— ln (c^{II}/c^{I}) = in (c^{I}/ c^{II})]. Thus, the Donnan potential can also be regarded as a type of diffusion potential occurring as the mobile ions tend to diffuse away from the charges of opposite sign, which remain fixed in the Donnan phase.

Active Transport

The expression "active transport" implies that energy derived from metabolic processes is somehow involved in moving a solute across a membrane toward a region of higher chemical potential. There are three different aspects to this description of active transport, viz., a supply of energy, movement, and an increase in chemical potential. Although not really necessary, the term "active transport" has conventionally been restricted to the case of movement in the energetically uphill direction, and we will follow this restriction here. By itself, a difference in chemical potential of a certain species across a membrane does not necessarily imply that active transport of that species is occurring. For example, if the solute does not penetrate the membrane, the solute is unable to attain equilibrium across it, and μ_j would not be expected to be the same on the two sides. For ions actually moving across some membrane, the ratio of the influx to the efflux of a particular ionic species provides information which helps to indicate whether or not active transport is taking place.

Specifically, the Ussing-Teroell, or flux ratio equation is satisfied for that species if it moves passively and independently across the membrane. A very simple and often effective approach for determining whether fluxes are active or passive is to remove possible energy sources. For photosynthesizing plant tissue, this can mean comparing the fluxes in the light with those in the dark. (In addition, we should check whether the permeability of the membrane changes or other alterations occur upon illumination). Compounds or treatments which disrupt metabolism also can be useful for ascertaining whether metabolic energy is being used for the active transport of various solutes. We will begin our discussion by indicating how active transport can directly affect membrane potentials.

We will then compare the temperature dependencies of metabolic reactions with those for diffusion processes across a barrier to show that a marked enhancement of solute influx caused by increasing the temperature does not necessary indicate that active transport is taking place. Next we will consider a much more reliable criterion for deciding whether fluxes are passive or not, viz., the Ussing-Teorell, or flux ratio equation. After examining a specific case in which active transport is involved, we will discuss a relationship that in often invoked for describing the concentration dependence of ion uptake.

Electrogenicity

One of the possible consequences of actively transporting a certain ionic species into a cell or organelle is the development of an excess of electrical charge inside the membrane surrounded entity. If for some ionic species the active transport process involves an accompanying ion of opposite charge or an equal release of a similarly charged ion, the total charge in the cell or organelle is unaffected. However, if the charge of the actively transported ions is not directly compensated for, the process is electrogenic, i.e., it tends to generate an electrical potential difference across the membrane.

An electrogenic uptake of an ion which gives a net transport of charge into the cell or organelle will thus affect the membrane potential. We can appreciate this effect by referring to Equation 1 ($Q=C\Delta E$), which indicates that the drop in potential across the membrane, ΔE, equals Q/C where Q is the net charge enclosed within the cell or organelle and C is its capacitance. The initial movement of net charge across the membrane by active transport leads to a fairly rapid charging or discharging of the membrane capacitance. To be specific, we will consider a spherical cell into which there is an electrogenic influx of chloride caused by active transport, $J_{a.t.cl}^{in}$. The amount of charge transported in time t across the surface of the sphere (area = $4\pi r^2$) is $J_{a.t.cl}^{in} 4\pi r^2 t$, where r is the radius. This active uptake of Cl^- increases the internal concentration of negative charge by the amount moved in divided by the cellular volume, or $J_{a.t.cl}^{in} 4\pi r^2 t/(4\pi r^3/3)$, which is $3J_{a.t.cl}^{in} t/r$. Let us suppose that $J_{a.t.cl}^{in}$ t is 1.0 picomole/cm^2 –sec (a picomole $=10^{-12}$ mole, and that the cell has a radius of 30 μm. In 1 sec the concentration of Cl^- actively transported in would then be $(3)(10^{-12}$ mole/cm^2 –sec)(1 sec)/(30 $\times 10^{-4}$ cm), or 1.0×10^{-9} mole/cm^3. Assuming a membrane capacitance of 1 μf/cm^2, we saw that such a cell has 1.0×10^{-9} mole of uncompensated negative charge/cm^3 when the interior is 100 mv negative with respect to the external solution. If no change were to take place in the other ionic fluxes, the electrogenic uptake of Cl^- into this cell would cause the interior to become more negative at the rate of 100 mv/sec.

The charging of the membrane capacitance is indeed a rapid process. The initiation of an electrogenic process causes an adjustment of the passive fluxes across the membrane. In particular, the net charge actively brought in is electrically compensated for very soon by appropriate passive movements of other ions into or out of the cell. The actual electrical potential difference across the membrane then results from the diffusion potential caused by these new passive fluxes plus a contribution form the electrogenic process involving the active transport of some charged species. We can represent the electrical potential drop generated by the active transport of species, $E_{a.t.j}$, as follows.

$$E_{a/t}j = z_j F J_{a.t.J} R^{memb} \quad ...(19)$$

where z_j is the charge number of species j, F is the Faraday, $J_{a.t.j}$ is the net flux of species j caused by active transport, and R^{memb} is the membrane resistance. $FJ_{a.t.j}$ in Equation 19—which is really a form of Ohm's law— has units of (coulombs/mole) × (moles/cm 2-sec) or coulombs/cm^2-sec, which is amperes/cm 2 . Hence, R ^{memb} must be expressed in ohms-cm^2, so that $zjFJ_{a.t.j}$ R^{memb} In Equation 19 can have units of ampere-ohms, i.e., volts the proper unit for electrical potential. For most plant cells, R ^{memb} is usually from 1000 to 8000 ohm-cm ^{2}.

Let us now see how large E $_{a.t.j}$ given by Equation 19 might be for representative electrogenic "pumps." The electrical potential drop created by active transport of 1 picomole of a monovalent cation/cm ^{2}-sec. Across a membrane resistance of 3000 ohm-cm^2 would be (+1) (96,487 coulombs/ mole) (10 $^{-12}$ mole/cm 2-sec) (3000 ohms-cm ^{2}), or 0.3×10^{-3} volt, which is a rather small potential difference. When the membrane-resistance is higher or the net flux by the electrogenic pump is greater, the potential created by the active transport process would of course be larger. For certain plant cells and some animal cells, R^{memb} is about 10,000 ohm-cm^2 and $j_{a.t.j}$ may be 20 picomoles/cm^2 –sec.

Equation 19 indicates that the potential drop created by such an electrogentic pump for a monovalent cation would be approximately (+1) (96,487)20 × 10^{-12})(10,000), or 0.019 v (19 mv). In any case, the actual electrical potential difference across a membrane is obtained by adding the potential drop caused by active transport of uncompensated charge (Equation 19) and that for passive fluxes, the latter being predicted by the Goldman equation (Equation 8).

Q_{10}, a Temperature Coefficient

In general, most metabolic reactions are markedly influenced by temperature, while processes like light absorption are insensitive to temperature. What temperature dependence should we except for diffusion? Can we decide whether the movement of some solute into a cell is by active transport or by passive diffusion once we know how the fluxes depend on temperature? To answer such questions, we need an expression describing the distribution of energy among the molecules as a function of temperature in order to determine what fraction of the molecules has the requisite energy for a particular process. In aqueous solutions, the relevant energy is generally the kinetic energy of motion of the species involved.

Hence, we will first relate the distribution of kinetic energy among molecules to temperature. At the end of this section we will present the related concept of activation energy—activation energies are used to discuss reactions which have an energy barrier that must be surmounted before the process can take place the topics. The topics that we will introduce here are important for a basic understanding of many aspects of biology— all the way from biochemical reactions to the consequences of light absorption. We begin by noting that very few molecules possess extremely large kinetic energies. In fact, the probability that a molecule has a given kinetic energy E decreases exponentially as E increases. The precise statement of this is known as the Boltzmann energy distribution, which describes the frequency with which specific kinetic energies are possessed by molecules by molecules at equilibrium at a given temperature. A convenient form for the Boltzmann energy distribution at temperature T is as follows:

$$n(E)=n_{total}\ e^{-E/kT} \qquad ...(20)$$

where n(E) is the number of molecules possessing an energy of E or more out of the total number of molecules, n_{total} and k is Boltzmann's constant k = 8.617 × 10 –5 electron volts (ev)/molecule-°K, where 1 ev is the energy change for a particle of unit electronic charge when it moves through an electrical potential difference of 1 volt. The quantity $e^{-E/kT}$, which equals n(E)/n total by Equation 20 is often referred to as the *Boltzmann factor*. Equation 20 indicates that the number of molecules with an energy of zero or greater, n(0), equals $n_{total}\ e^{-0/kT}$ or $n_{total}\ e^{-0}$, which is simply n total, the total number of molecules present. Due to collisions based on thermal motion, energy is continually being gained or lost by individual molecules in a random fashion.

Hence a wide range of kinetic energies is possible, although very high energies are less probable, of Equation 20. As the temperature is raised, not only is the average energy per molecule higher, but the number of molecules in the "high energy tail" of the exponential Boltzmann distribution also increases substantially. For diffusion across a membrane, the appropriate Boltzmann energy distribution indicates that the number of molecules with a kinetic energy of U or greater per molecules resulting from velocities in some particular direction is proportional to $\sqrt{T}e^{-U/kT}$.

A minimum kinetic energy (U_B) is often necessary to diffuse past some barrier or to cause some specific reaction. Thus, any molecule with a kinetic energy of U_B or greater has sufficient energy for the particular process. For the Boltzmann energy distribution appropriate to this case, the number of such molecules is proportional to $\sqrt{T}\ e^{-U_B/KT}$ (These expressions having the factor $\sqrt{T}$ actually only apply to one-dimensional cases, e.g., for molecules diffusing across a membrane). At a temperature 10° higher, the number is proportional to $\sqrt{(T+10)}e^{-}\ U_B/[K(T+10)]$. The ratio of these two quantities is called the Q_{10}, or temperature coefficient of the process:

$$Q_{10} = \sqrt{\frac{T+10}{T}}\,\frac{10U_B}{e^{kT(T+10)}} \qquad \text{...(21)}$$

To obtain the form of the exponential given in Equation 21, we note that $U_{B/[k(T+10)]+UB/(kT)}$ is equal to $TU_{B/[kT(T+10)]} + (T+10)U_{B/kT(T+10)]}$

A Q_{10} near one is characteristic of some passive processes which have no energy barrier to surmount, i.e., where U_B equals 0. On the other hand, most enzymatic reactions take place only when the reactants have a considerable kinetic energy, so that such processes tend to be quite sensitive to temperature. A value of 2 or greater for Q_{10} is often considered to indicate the involvement of metabolism, such as occurs for active transport of a solute into a cell or organelle. However, Equation 21 indicates that any process having an appreciable energy barrier (U_B) can have a large temperature coefficient. The value of the temperature coefficient Q_{10} for a particular process can be indicative of the minimum kinetic energy required (U_B), and vice versa. A membrane often represents an appreciable energy barrier for the diffusion of charged solutes; let us say that the U_B for passive ion movement across it is 0.50 ev/molecule.

Using Equation 21, this would lead to a temperature coefficient for T equalling 20° of $\sqrt{(303)/293}\ e^{(10)(0.50)/(8.617\times10^{-5})(293)(303)}$, or $1.02\ e^{0.654}$, which equals 1.96. In other words, the uptake of this particular ion would approximately double with only a 10° increase in temperature. This example indicates that a passive process can have a rather high Q_{10} if there is an appreciable energy barrier, indicating that a large temperature coefficient for ion uptake does not necessarily imply active transport. A kinetic energy of 0.50 ev or greater is possessed by only a relatively few molecules in the Boltzmann energy distribution—$n(E)=n_{total}\ e^{-E/kT}$, Equation 20. For instance, at 20° the Boltzmann factor $e^{-E/kT}$ is $e^{-(0.50\ ev)/(0.0753\ ev)}$, or only 2.5×10^{-9}(kT = 0.0253 ev at 20°. As T is raised, the number of such molecules in the high energy part of the Boltzmann distribution increases greatly. Many more molecules thus have the requisite kinetic energy, U_B, and consequently can make part in the process being considered.

In particular, at 30° the Boltzmann factor in Equation 20 becomes e^{-} (0.50)/(0.261), or 4.6×10^{-9} for a kinetic energy of 0.50 ev/molecule (kl=0.0261 ev at 30°). Hence, the Boltzmann factor for a U_B of 0.50 ev/molecule essentially doubles for a 10° rise in temperature, consistent with our Q_{10} calculation for this case. So far in this section we have calculated on the basis of energy per molecule, while in many cases it is more convenient to consider energy per mole. In order to change from a molecule to a mole basis, we multiply Boltzmann's constant k(energy/

molecule-°K) by Avogadro's number N (molecules/mole), which gives us the gas constant R energy/mole-°K) i.e., R equals kN. If U and U_B are expressed as energy per mole, we simply replace k in the Boltzmann energy distribution (Equation 20) and Equation 21 by R. Finally, as indicates, l ev/molecule is equal to 23.06 kcal/mole. The existence of an energy barrier of height U_B is related to the concept of activation energy, which refers to the minimum amount of energy necessary for some reaction to take place.

In the case of a membrane, we can evaluate U_B by determining how the number of molecules diffusing across the membrane varies with temperature, e.g., by invoking Equation 21. In fact the kinetic energy (U_B) corresponds to the activation energy for crossing the barrier. For a chemical reaction, we can also experimentally determine how the process is influenced by temperature. If we represent the activation energy per mole by A, the rate constant for such a reaction should vary with temperature as follows:

$$\text{Rate constant} = Be^{-A/RT}$$

where B is essentially a constant.

Equation 22 is referred to as the *Arrhenius equation.* It was originally proposed on experimental grounds by Arrhenius at the end of the 19th century and only subsequently interpreted theoretically. A plot of the logarithm of the rate constant versus 1/T is commonly known as an Arrhenius plot. Hence, the slope of an Arrhenius plot is proportional to the activation energy by Equation 22–ln (rate constant)=ln B-A/RT. Many reactions of importance in biochemistry have large values for A and are therefore extremely sensitive to temperature. Moreover, an enzyme greatly increases the rate of the reaction which it catalyzes by reducing the value of the activation energy needed (Equation 22). Thus, the related concepts of activation energy and energy barrier have many applications in biology. Let us finally consider the activation energies for certain cases of diffusion. For diffusion in water at 20°, A is about 4 kcal/mole for K^+ and 5 kcal/mole for mannitol (see Stein). After replacing U_B by A and k by R in Equation 21, we can calculate that the appropriate Q_{10} for such diffusion would be 1.3. The activation energy for the passive efflux of K + from many cells is about 15 kcal/mole near 20°, corresponds to a Q_{10} of 2.5. Again, we conclude that a purely passive process, such as diffusion across membranes, can have a marked temperature dependence.

Ussing-Teorell Equation

We will now consider ways of distinguishing between active and passive fluxes which are more reliable than determining Q_{10}s. One of the most useful physicochemical criteria for deciding whether a particular ionic movement across a membrane is active or passive is the application of the *Ussing-Teorell*, or flux ratio equation. For ions moving passively, this expression shows how the ratio of the influx to the efflux depends on the internal and external concentrations of that species as well as on the electrical potential difference across the membrane. If the Ussing-Teorell equation is satisfied, active transport of the ions need not be invoked. We can readily derive this expression by considering how the influx and efflux could each be determined experimentally, as the following arguments indicate. For measuring the unidirectional inward component, or influx of a certain ion Jj^{in}, the plant cell or tissue can be placed in a solution containing a radioactive isotope of species j.

Initially, none of the radioisotope is inside the cells, hence internal specific activity for this isotope equals zero at the beginning of the experiment. (As for any radioisotope study, only some of the molecules of species j are radioactive. The particular fraction is known as the *specific activity*, and must be determined for both cj° and cj^i in the present experiment.) Since none is inside, the initial unidirectional outward component or efflux of the radioisotope, Jj^{out}, is zero, therefore, the initial net flux (Jj) of the isotope reflects Jj^{in}. From Equation 14, this influx of the radioisotope can be represented by (KjujzjFE M/Δx) [1/(e tj FEM/RT –1)]CJ°. After the isotope has entered the cell, some of it will start coming out. Therefore, only the initial flux will give an accurate measure of the influx of the radioisotope of species j. Once the radioactivity has built up inside to a substantial level, we remove the radioisotope from the outside solution. The flux of the isotope will then be from inside the cells to the external solution. In this case, the specific activity for cj° equals zero, and cj^i determines the net flux of the radioisotope.

From Equation 14, this efflux of the isotope of species *j* differs in magnitude from the initial influx only by having the factor cj° replaced by $cj^i e^{ij\,FEM/RT}$, while the quantities in the first tv'o parentheses remain the same. The ratio of these two fluxes —each of which can be separately measured ¾takes on the following relatively simple form :

$$\frac{J_j^{in}}{J_j^{out}} = \frac{cj^{\circ}}{cj^i e^{x/FE_M/RT}} \qquad ...(23)$$

Equation 23 was independently derived by both Ussing and Teorell in 1949, and is known either as the Ussing-Teorell equation or the flux ratio equation. It is strictly valid only for ions moving passively without interacting with other substances that may also be moving across the membrane. The present derivation makes use of Equation 14, which gives the passive flux of some charged species across a membrane in response to differences in the chemical potential of that species across the barrier. Equation 14 was derived by considering only one species at a time, hence, possible interactions between the fluxes of different species are not included in Equation 23.

The Ussing-Teorell equation can thus be used to determine whether the observed influxes and effluxes are passive (i.e., responses to the chemical potentials of the ions on the two sides of a membrane), or whether additional factors such as interactions between species or active transport must be invoked. For example when active transport of species *j* into the cell is taking place, the actual J_j^{in} would be the passive unidirectional flux (i.e., the one predicted by Equation 23 *plus* the influx due to active transport. After one further comment about Equation 23, we will consider a specific example of the use of the Ussing-Teorell equation to evaluate membrane fluxes. The ratio of the influx of species *j* to its efflux, as given by Equation 23, can readily be related to the difference in its chemical potential on the two sides of the membrane, $\mu j^{\circ}—\mu j^{i}$. This difference causes the flux ratio to differ from unity, and we will return to it to estimate the minimum amount of energy needed to actively transport that ionic species across the membrane. After taking logarithms of both sides of the Ussing-Teorell equation (Equation 23) and multiplying by RT, we obtain the following equalities:

$$RT \ln \frac{J_j^{in}}{J_j^{out}} = RT \ln \frac{cj^{\circ}}{cj^{i}} - zjFE_M$$

$$= RT \ln aj^{\circ} + zjFE^{\circ} - RT \ln aj^{i} - zjFE^{i}$$

$$= \mu j^{\circ} - \mu j^{i} \quad ...(24)$$

where the membrane potential (E_M) has been replaced by E^i - E° in keeping with our previous convention. The derivation is restricted to the case of constant $\gamma j(\gamma^{\circ}=\gamma j^{i})$, which means that In ($cj^{\circ}/cj^{i}$) equals ln ($\gamma j^{\circ}\, cj^{\circ}/\gamma j^{i}\, cj^{i}$), or ln ($aj^{\circ}/aj^{i}$), which is in aj°- In aj^{i} . Finally, the 'VjP term in the chemical potential is ignored for the charged species - actually, we need only assume that 'VjP° so that μj then equals μj^{*} + RT In aj + zjFE, where μj^{*} has the same value on the two sides of the membrane. Thus, $\mu j^{\circ} - \mu j^{i}$ is RT ln aj°+zjFE° - RT In aj^{i} - zjFE ^{i} , as is indicated in Equation 24.

A difference in chemical potential of species *j* across a membrane would cause the flux ratio of the passive fluxes to differ from unity, a conclusion that follows directly from Equation 24. When μj° equals μj^{i} the influx would balance the efflux, hence there would be no net passive flux of species *j* across the membrane (Jj=Jj° ^{out} by Equation 14). This condition (μj°=μj^{i}) is also described by Equation 4, used to derive the Nernst equation. In fact, in such a case of equality of the chemical potentials, the electrical potential difference across the membrane is the Nernst potential, E_N, given by Equation 7, Enj=(RT/$_z$j F) In (aj°/aj ^{i}). Thus, when E_M equals E_Nj for some species *Jj* ^{In} equals J*j* out, and no net passive flux of that ion is expected cross the membrane nor is any energy expended in moving the ion from one side of the membrane to the other.

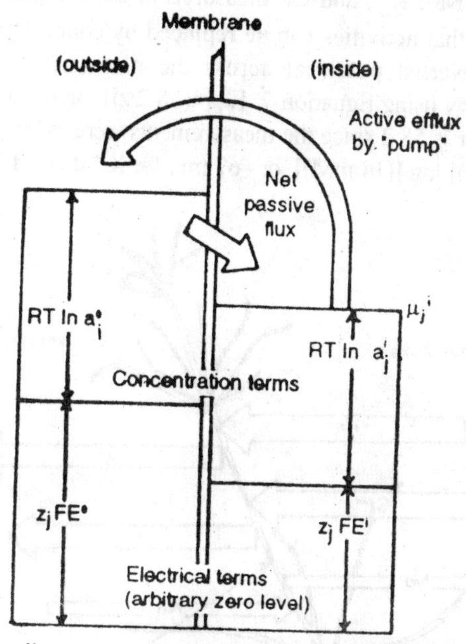

Fig. 4.4. Diagram illustrating the situation for an ion not in equilibrium across a membrane. Since μ_j^0 is greater than μ_j^i there is a net passive flux into the cell.

When Equation 23 and 24 are not satisfied for some species, such ions are not moving across the membrane passively, or perhaps not moving independently from other fluxes. One way this may occur is for the various fluxes to be interdependent, a condition describable by irreversible thermodynamics. Another way is through active transport of the ions, whereby energy derived from metabolism is used to move solutes to regions of higher chemical potential.

Example of active transport

The above principles for deciding whether or not active transport of certain ions is taking place can be illustrated by using data obtained with the large internodal cells of *Nitella translucens*. All the parameters appearing in the Ussing-Teorell equation have been measured for Na^+, K^+, and Cl^- using *Nitella*. For experimental purposes, this fresh water alga is often placed in a dilute aqueous solution containing 1 mM NaCl, 0.1 mM KCl plus 0.1 mM $CaCl_2$—this solution is not unlike the pond water in which Nitella grows, and is generally referred to as "artificial pond water" – which indicates the values for all three cj°s. The concentrations Na^+, K^+, and Cl^- measured in the cytoplasm.

Assuming that activities can be replaced by concentrations, we can calculate the Nernst potential across the plasmalemma from these concentrations by using Equation 7. $E_N j = (58.2/zj) \log (cj^o/cj^i)$ in mv (the numerical factor is 58.2 since the measurements were at 20°). $E_N j$ for Na^+ is [(58.2 mv)/(1)] log [(14 mM)], or –67 mv; for K^+ it is – 179 mv; and for

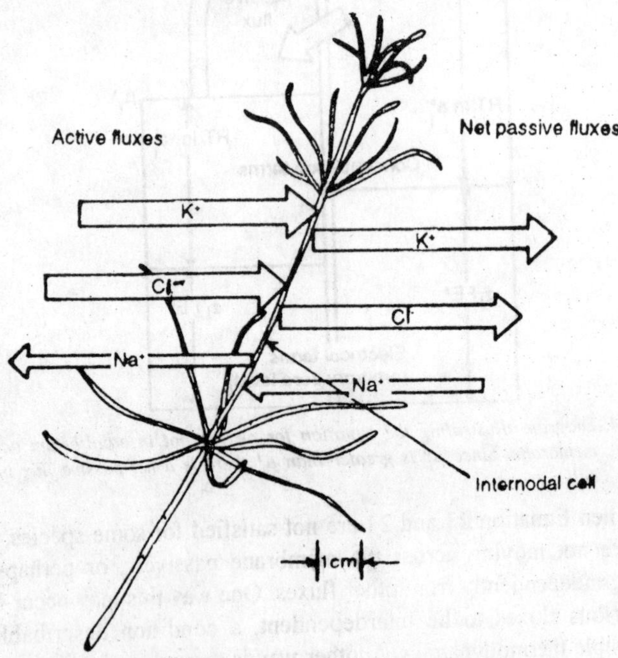

Fig. 4.5. In the steady state condition in the light, the three active fluxes across the plasmalemma of the large internodal cell of Nitella are balanced by the net passive effluexes of potassium and chloride and the net passive influx of sodium.

Cl⁻ the Nernst potential is 99 mv. A direct measurement of the electrical potential drop across the plasmalemma (E_M) gives −138 mv, as was indicated earlier in discussing the Goldman equation. Since E_{Nj} differs from E_M in all three cases, none of these ions is in equilibrium across the plasmalemma of *Nitella*. A difference between E_M and E_{Nj} for a particular species indicates departure from equilibrium for that ionic species; it also tells in what compartment μj is higher. Specifically, if the actual membrane potential is algebraically more negative than the calculated Nernst potential, the chemical potential in the inner aqueous phase (here the cytoplasm) is lower for a cation but higher for an anion, compared to the values in the external solution—cf. The effect of zj in the electrical term in the chemical potential, zjFE. Since E_M (−138 mv) is more negative than E_{NNn} (−67 mv), is at a lower chemical potential in the cytoplasm than outside in the external solution.

Reasoning in an analogous manner, we find that K^+ (E_{NK} = −179 mv) has a higher chemical potential inside, while Cl^- (E_{NCl} = 99 mv) is at a much higher chemical potential inside. If these ions can move across the plasmalemma, this suggests an active transport of K^+ and Cl^- into the cell and an active extrusion of Na^+ from the cell. We can also consider the movement of the various ions into and out of Nitella in terms of the Ussing –Teorell equation to help determine whether active transport need be invoked to explain the fluxes. The Ussing-Teorell equation predicts that quantity appearing on the right side of Equation 23 ($cj°/cj\ i\ e^{2jFEM/RT}$) should equal the ratio of influx to efflux of the various icons, if the ions are moving passively in response to gradients in their chemical potential. Using the value given in Table 4.4, this ratio (*Jj* in/*Jj* out) is calculated for sodium as follows: the factor RT/F is 25.3 mv at 20° hence $e^{tjEFM/RT}$ is $e^{(1)(-138}$ mv)/(25.3 ^{mv}) or $e^{-5.45}$ for an E_M of 0138 mv. Therefore, c_{Na} °/(CNa'e $_{-zNaFEM/RT}$) is (1.0 mM)/(14 mM)(e −5.45)], which equals 17. Following the same procedure, the expected flux ratio is calculated to be 0.20 for K + and 0.000085 for Cl^-. However, the observed influxes in the light equal the effluxes for each of these three ions.

Equal influxes and effluxes are quite reasonable for mature cells of Nitella, which can be described by a steady-state condition. On the other hand, if Jj^{in} equals, the Jj out, the flux ratios given by Equation 23 are not satisfied for Na^+, K^+, or Cl^-. Active transport of K^+ and Cl^- in and Na^+ out accounts for the marked deviations from the Ussing-Teorell equation for *Nitella*. As we mentioned earlier, another approach for studying active transport is to remove the supply of energy. In the case of *Nitella*, cessation of illumination causes an appreciable decrease in the Na efflux, in the K influx, and in the Cl influx. But these are the three fluxes which are toward regions of higher chemical potential for the particular ions involved, and thus we may reasonably expect all three to be active. On the other hand, some fluxes remain essentially unchanged by placing the cells in the dark (refer to the fluxes soon after extinguishing the light, not the steady-state fluxes).

For instance, the Na influx and K efflux are initially unchanged when the Nitella cells are transferred from the light to the dark, i.e., these unidirectional fluxes toward lower chemical potentials do not depend on energy derived from photosynthesis. (For reasons which are as yet unclear, the energetically downhill efflux of Cl^- apparently increases in the dark.) The passive diffusion of the ions toward regions of lower chemical potential helps create the electrical potential difference across a membrane, while active transport is instrumental in maintaining the asymmetrical ionic distributions which sustain the passive and the active fluxes are interdependent in the ionic relations of cells and both are crucial for the generation of the observed diffusion potentials.

Moreover, active and passive fluxes can occur simultaneously in the same direction. For example, we calculated that J Na in should equal 17 times J_{Na} out, if both fluxes were passive ones obeying the Ussing-Teorell equation. Since J_{Na} in is passive and equal to 0.55 picomole/cm^2 –sec, we expect a passive efflux of Na + equaling (0.55)/(17), or 0.03 picomole/cm^2 –sec. Thus, the active component of the Na efflux in the light may be 0.55–0.03, or 0.52 picomole/cm^2 –sec. At cessation of illumination, J_{Na}^{out} decreases from 0.55 to 0.1 picomole/cm^2 –sec. Extinguishing the light removes photosynthesis as a possible energy source for active transport, but respiration could still supply energy in the dark. This helps explain why J_{Na}^{out} in the dark does not decrease all the way to 0.03 picomole/cm^2 -sec, the value predicted for the passive efflux. Suppose that the chemical potential some species is higher outside than inside a cell.

The minimum amount of energy needed to transport a mole of some membrane species from the internal aqueous phase on one side of some membrane to the external solution on the other is the difference in chemical potential of that solute across the membrane, $\mu j^\circ - \mu j^i$ there, $\mu j^\circ > \mu j^i$). As we noted in considering Equation 24, the quantity μj° - μ^i for ions is RT In (aj°/aj^i) -zjFE_M. Since the Nernst potential, $E_N j$, is (RT/zjF) In (aj°/aj^i) (Equation 7). We can express the difference in chemical potential across the membrane as follows:

$$\mu j^\circ - \mu j^i = zjF(E_{Nj} - E_M)$$
$$\mu j^i - \mu j^\circ = zjF(E_M - E_{N}j) \quad ...(25)$$

We will use the bottom line of Equation 25 to discuss the active transport of ions into cells (i.e., when $\mu j^i > \mu j^\circ$).

Using the nernst potentials of Na^+, K^+, and Cl^- for Nitella translucens and the value of E_M, we can calculate $zj(E_M—E_Nj)$ in Equation 25 for transporting these ions across the plasmalemma of this alga. Such a quantity is (+1)(–138 mv)–(–67mv), or–71 mv for Na^+,+41 mv for K^+, and +237 mv for Cl^- . These values for $zj(E_M–E_{Nj})$ mean that Na^+ is at a higher chemical potential in the external bathing solution while K^+ and Cl^- are at higher chemical potentials inside the cell, as we concluded above. From Equation 25 the minimum energy required to actively transport, or "pump" Na^+ out across the plasmalemma of the *Nitella* cell equals F times the energy difference in millivolts, which is (0.02306 kcs/mole-mv) (71 mv), or 1.6 kcal/mole in the case of Na^+ (F is 0.02306 kcal/mole-mv). Potassium requires (0.02306) (41), or 0.9 kcal/mole to be pumped inward. The active extrusion of Na^+ from certain algal cells may be linked to the active uptake of K^+, ATP being implicated as the energy source for this coupled exchange process.

The hydrolysis of ATP under biological conditions usually releases at least 10 kcal of energy/mole. For the case of Nitella cell, this is sufficient energy per mole of ATP hydrolyzed to pump a mole of Na^+ out (1.6 kcal/mole) and one of K^+ in (0.9 kcal/mole). The transport Cl^- inward takes a minimum of (0.02306)(237), or 5.5 kcal/mole, according to Equation 25 which is fairly large amount of energy. Neither the form of the energy used nor the actual mechanism in actively transporting chloride into Nitella or other plant cells in fully understood at the present time, although exchanges with OH appear to be involved. The active uptake of K^+ and Cl^- together with an active extrusion of Na^+ in the case of *Nitella* are actually fairly widespread phenomena among plant cells. We might ask, why does a cell actively transport K^+ and Cl^- in and Na^+ out? Although no completely satisfactory answer can be given to such question, we shall *speculate* on possible reasons, based on the principles we have been considering.

Let us imagine that a membrane-bounded cell containing negatively charged proteins is placed in an NaCl solution, the latter possibly reflecting primeval conditions when life originated. When Na^+ and Cl^- are both in equilibrium across the membrane, E_M equals E_{Nct} and EN_{Na}. Using concentrations in Equation 3,5, log $(CN_a{}^o/ CN_a{}^i)$ then equals – log $(CC_l{}^o/CC_l{}^i)$, or $CN_a{}^o/CN_a{}^i = CC_l{}^i/CC_t{}^o$, hence $CN_a{}^oCC_l{}^o = CN_a{}^iCC_l{}^i$. For electroneutrality in an external solution containing only NaCl, $CN_a{}^o$ equals CCl° Since $a^2 = bc$ implies that $2a \leq b+c$, we conclude that $CN_a{}^o + CC_l{}^o \leq CN_a{}^i + CC_l{}^i$. But the proteins which cannot diffuse across the membrane, also make a contribution to the internal osmotic pressure, so that π^o would be less than π^o. Water tends to flow of course toward higher osmotic pressures. When placed in an NaCl solution, water would therefore tend to enter such a membrane-bounded cell containing proteins causing it to swell without limit. Thus, an outwardly directed active transport of Na^+ might have originated to lower CN_a^i and thus to prevent excessive swelling of primitive cells.

An energy-dependent uptake into a plant cell tends to increase π^i, which leads to a rise in P^i. This higher internal hydrostatic pressure favours cell enlargement and consequently cell growth. For a plant cell surrounded by a cell wall we might therefore expect an active transport of some species into the cell, e.g., chloride. (Animal cells do not have to push against a cell wall in order to enlarge, and do not generally have an active uptake of Cl^- Enzymes have evolved that operate efficiently when expected to relatively high concentrations of K^+ and Cl^-. The fact, many actually require K^+ for their activity, so that an inwardly directed K pump is probably necessary for biochemistry as we know it. The presence of a substantial concentration of such ions insures that electrostatic effects adjacent to a Donnan phase could in large part be screened out – otherwise, a negatively charged substrate e.g., an organic acid or a phosphorylated sugar, might be electrostatically repelled from the catalytic site on an enzymes (proteins are generally negatively charged at cytoplasmic pH's). Once active transport has set up certain concentration differences across the membrane, the membrane potential is an inevitable consequence of the tendency of such ions to passively diffuse toward regions of lower chemical potential. Whether such diffusion potentials actually benefit the general plant cell is an open question. They are essential however for the transmission of electrical impulses in excitable cells in animals and in certain plants.

Carriers–solute uptake and the Michaelis-Menten Formalism

Although active transport is clearly of common occurrence its actual mechanism – including the means whereby metabolic energy is used– remain uncertain. The possible involvement of a "carrier" molecule in such active transport of solutes across plant cell membranes was first suggested by W.J.V.Osterhout in the early 1930s. This carrier was proposed to bind selectively certain molecules and then to carry, or ferry them across the membrane. Carriers can provide a cell with the specificity or selectivity needed to control the entry and exit of the various types of solutes encountered. Thus, certain metabolites can be specifically taken into the cell, while photosynthetic and waste products can be selectively moved out across the membranes. Also, the active transport of certain inorganic nutrients into epidermal cells in the root allows a plant to obtain and accumulate these solutes from the soil.

Even though the all-important mechanism for physically binding and moving the solutes through the membrane is still unknown, the carrier concept has found widespread application in the interpretation of experimental observations. Based primarily on results from competition studies, certain solutes are described as being bound to, or associated with a particular carrier. When an ion of some species is attached to a carrier, another similar ion (of the same or a different species) competing for the same binding site cannot also be bound. For example, the similar monovalent cations K^+ and Rb^+ appear to bind in competitive fashion to the same carries. For some cells, one and the same carrier might transport Na^+ out of the cell and K^+ in – the so-called sodium-potassium pump alluded to above.

Calcium and strontium may compete with each other for binding sites on another common carrier. Two other divalent cations, Mg^{++} and Mg^{++}. are apparently transported by a single carrier that is different from the one for Ca^{++} and Sr^{++}. The halides (Cl^- I^-, and Br^-) may also be transported by a single carrier. Although ions have been shown to bind specifically to certain proteins isolated from cellular membranes, much remains to be learned about carriers.

One of the most important variables in the study of carrier mediated uptake is the external concentration. As the external concentration of a solute increases, the rate of uptake generally reaches an upper-limiting value. We may then presume that all binding sites on the carriers for that particular ion or neutral solute have become filled or saturated. In particular, the rate of active uptake of species j, Jj^{in}, is often proportional to the external concentration of that solute ($cj°$) over the lower range of concentrations, leveling off as $cj°$ is raised, and eventually approaching a maximum rate for the influx ($Jj^{in}{}_{max}$). We can describe this kind of behaviour by the following equation :

$$J_j^{in} = \frac{J_{j\,max}^{in}\, c_j^o}{K_j + c_j^o} \qquad ...(26)$$

where Kj is a constant characteristic of species *j* crossing a particular membrane, expressed in the units of concentration. For the uptake of many ions into roots and other plant tissues, $Jj^{in}{}_{max}$ is 1 to 10 μ moles/g fresh weight-hr. Often, two different *Kj*s are observed for the uptake of one and the same ion into a root. The lower one is generally between 6 and 100μM (which is the concentration range of many ions in the soil water), while the other Kj can be above 10 mM.

Equation 26 is similar in appearance to the Michaelis-Menten equation used for describing enzyme kinetics in biochemistry. The substrate concentration in the latter relation is analogous to c *j* in Equation 26 and the enzyme reaction velocity to J_j^{in}. The term in the Michaelis-Menten equation equivalent to K *j* in Equation 26 is the substrate concentration for half-maximal velocity of the reaction, K_M (the Michaelis constant). We can press the analogy between enzyme kinetics and active uptake even further. The lower the K_M, the greater is the affinity of the enzyme for the substrate. Likewise a low value for K *j* indicates that the ion or other solute is more readily bound to some substance in the membrane such as a "carrier" molecule.

The physiological consequence of a low K *j* is that species j can be efficiently bound, even when its external concentration is relatively low; such a species is actually favored or selected for active transport into the cell. The two most common ways of graphing data on solute uptake which fit Equation 26 shows that when the external concentration of species j, cj°, is equal to K_j , Jj is equal to Kj, J*j* in equals ½$J_j^{in}{}_{max}$ as we can see directly from Equation 26 I.e.J_j^{in} is then $J_j^{in}{}_{max}$ k/K*j*+Kj), or ½$J_j^{in}{}_{max}$. Thus, K*j* is the external concentration at which the rate of active uptake is half-maximal – in fact, the observed values of Kj have proved to be convenient parameters for describing the uptake of various solutes. If two different solutes compete for the same site on some carrier, then J_j^{in} for species *j* will be decreased by the presence of the second solute, which is known as *competitive inhibition* of species *j*.

As the concentration cj° is increased, the uptake continuously rise and eventually approaches an upper limit, $J_j^{in}{}_{max}$, which is the asymptotic value for the active influx. In the case of competitive inhibition, $Jj^{in}{}_{max}$ is not affected, since in principle we can raise cj° high enough so that the same maximum rate for the active uptake of species *j* is eventually obtained. However, the half-maximum rate occurs at a higher concentration, which is another way of saying that Kj is raised if a competing solute is present. For many purposes, it is advantageous to plot the experimental data in such a way that a linear relationship is obtained when Equation 26 is satisfied for the active uptake of species j.

Taking reciprocals of both sides of Equation 26 we note that I/J_j^{in} equals $(K_j + c_j°)/(J_j^{in}{}_{max}$ cj°), which is Kj/J^{in}_{max} cj°) + I/$J_j^{in}{}_{max}$. Hence, when I/Jj in is plotted versus I*/cj°, Equation 26 takes the form of a straight line of slope Kj/$Jj^{in}{}_{max}$, with an intercept on the ordinate of I/$J_j^{in}{}_{max}$. This later method of treating the experimental results has proved to be rather convenient, and has found widespread application in studies of solute uptake by plant tissues, especially into roots. Since Kj increases while Jj^{in}_{max} remains the same for competitive inhibition, the presence of a competing species causes the slope of the line (i.e., Kj/Jj^{in}_{max}) to be greater, while the intercept on the y-axis (I/Jj^{in}_{max}) is unchanged. Certain compounds that are structurally unrelated to species j can inhibit the functioning of the carriers. Such a noncompetitive inhibitor does not bind to the site used for transporting species j across the membrane, and Kj is unaffected. Since a noncompetitive inhibitor lowers the maximum rate of active influx (Jj^{in}_{max}), both the slope and the y-axis intercept changed. The basis of the mathematical form of Equation 26 for describing uptake is the competitive binding of solutes to specific sites on a limited number of carriers.

In other words, active processes involving metabolic energy do not necessarily have to be invoked. If a solute were to diffuse across the membrane only when bound to a carrier, the expression for the influx could also be Equation 26. This passive entry of a solute mediated by a carrier is termed *facilitated diffusion.* Since facilitated diffusion is so important in biology and yet is often misunderstood, we will briefly elaborate upon it here. Certain molecules passively enter cells more readily than would be expected from consideration of their molecular structure or observations with analogous substances (e.g. by comparing Pjs). It would appear that some mechanism is facilitating their entry. Moreover, the net flux is still toward lower chemical potentials and hence would be in the same direction as ordinary diffusion. To help explain facilitated diffusion, carriers are proposed to act as shuttles for a net passive movement of the specific molecules across the membrane towards, regions of lover energy.

Instead of the usual diffusion across the barrier—based on random thermal motion of the solutes—carriers select out and bind certain molecules. The solute-carrier complex then diffuses across the membrane—or perhaps the solute-carrier complex simply rotates 180° in the membrane— and, finally, the solutes are somehow released on the other side of the membrane without the intervention of metabolism. Such facilitation of entry by carriers may also be regarded as a special means of lowering the activation energy needed for the solute to cross the energy barrier represented by the membrane. Thus, carriers facilitate the influx of solutes in the same way that enzymes facilitate biochemical reactions. There are certain general characteristics of a facilitated diffusion system.

As already mentioned, the net flux is toward lower chemical potentials. (According to the usual definition, active transport is in the energetically uphill direction and, conceivably, it can employ the same carriers used in facilitated diffusion.) Movement by facilitated diffusion is more rapid than that expected for ordinary diffusion. Also, the carriers can exhibit a high degree of selectivity, i.e., they can be specific for certain types of molecules, while not binding closely related species. In addition carriers in facilitated diffusion become saturated when the external concentration of the species, transported is raised sufficiently, a behaviour consistent with Equation 26.

Finally, since carriers can exhibit competition phenomena, the flux of a species entering the cell by facilitated diffusion can be reduced when structurally and analogous molecules are added to the external solution. (Such molecules would compete for sites on the carrier and thereby reduce the binding and subsequent transfer of the original species into the cell.) For convenience, we have been discussing facilitated diffusion into a cell, while exactly the same principles apply for exit. Let us assume that a carrier for potassium exists in the membrane of a certain cell and that it is being used as a shuttle for facilitated diffusion. Not only would the carrier lead to an enhanced net flux toward the side with the lower chemical potential, but also both the unidirectional fluxes Jk^{in} and Kk^{out} could be increased over the value predicted for ordinary diffusion. This increase in the unidirectional fluxes by a carrier has often been called "exchange diffusion". In such a case the molecules are interacting with a membrane component, viz. the carrier, hence the Using-Teorell equation—Equation 23 $Jj\ in/Jj\ ^{out}=cj^{\circ}/(cje\ ^{zjFE\ M/RT})$—would not be obeyed, since it pre-supposes that there are no interactions with other substances.

In fact, observation of departures from predictions of the Using Teorell equation is often the way cases of exchange diffusions are actually discovered. Both active and passive fluxes across the cellular membranes can occur concomitantly; these movements depend on concentrations in rather different ways. For passive diffusion, the unidirectional component Jj^{in} is proportional to cj° for neutral solutes $[Jj=Pj(cj^{\circ}-cj^{i})]$, and by Equation 14 for ions. This proportionality strictly applies only over the range of external concentrations for which the permeability coefficient is essentially independent of concentration, and the membrane potential must not change in the case of charged solutes. Nevertheless, ordinary passive influxes do tend to be proportional to the external concentration, while an active influx or the special passive influx known as facilitated diffusion—either of which can be described by a Michaelis-Menten type of formalism—shows saturation effects at the higher concentrations. Moreover, facilitated diffusion and active transport exhibit both a high degree of selectively and competition phenomena, while ordinary diffusion does not.

PRINCIPLES OF IRREVERSIBLE THERMODYNAMICS

So far, we have been using classical thermodynamics—though it may have been noticed that we have often done so somewhat illegitimately. For example, let us consider Equation (Jvw=LwΔψ). Since there is a difference in chemical potential given by a change in water potential, we expect a net (irreversible) flow of water from one region to another, which is obviously not an equilibrium not an equilibrium situation. Strictly speaking, classical thermodynamics is concerned solely with equilibria, not with movement—classical thermodynamics might have been better-named "thermostatics". Thus, we have frequently been conc erned with a kind of hybrid—appealing to classical thermodynamics for the driving forces, and using non-thermodynamic arguments and analogies to discuss fluxes.

One of the objectives of irreversible thermodynamics is to help legalize the arguments. But, as well see, legalizing them brings in new ideas and considerations. Irreversible thermodynamics uses the same parameters as classical thermodynamics, viz., temperature, pressure, free energy, and so on. But these quantities have been strictly defined for macroscopic amounts of matter in equilibrium situation only. How can we use them to discuss processes *not* in equilibrium, the domain of irreversible thermodynamics? This dilemma immediately circumscribes the range of validity of the theory of irreversible thermodynamics it can only deal with "slow" processes or situations which are not very far from equilibrium, for only in such circumstances can equilibrium related concepts such as temperature and free energy retain their validity—at least approximately.

We have to assume from the outset that we can talk about and use classical thermodynamic parameters even in non-equilibrium situations. Another refinement that we should introduce into our theory is to recognize that the movement of one species may affect the flux of a second species. A particular flow of some solute may interact with another flux by way of collusions, each species flowing under the influence of its own force. For example, water, ions, and other solutes moving through a membrane toward regions of lower values for their respective chemical potentials can exert a frictional drag on each other. In this case, the magnitude of the flus of a solute may depend on whether water is also flowing. In this way, the fluxes of various species across a membrane become interdependent. Stated more formally, the flux of a solute is not only dependent on the negative gradient of its own chemical potential—which is the sole driving force we have recognized up to now—but it may also be influenced by the gradient in the chemical potential of water.

Again using Equation ($Jvw = Lw\Delta\psi$) as an example, we have considered that the flow of water depends only on the difference in its own chemical potential between two locations, and have thus far ignored any coupling to concomitant fluxes of solutes. A quantitative description of interdependent fluxes and forces is given by irreversible thermodynamics, a subject which treats non-equilibrium situations such as those actually occurring under biological conditions. In this brief introduction to irreversible thermodynamics we will emphasize certain underlying principles and then derive the reflection coefficient, a parameter important in plant and cell physiology. In order to keep the analysis manageable, we will restrict our attention to isothermal conditions, which approximate many biological situations where fluxes of water and solutes are considered.

Fluxes, Forces, and Onsager Coefficients

In our previous discussion of fluxes, the driving force leading to the flux of species *j*, *Jj*, was the negative gradient in its chemical potential, $-\partial\mu j/\partial x$. Irreversible thermodynamics takes a more general view, namely, that the flux of species *j* depends not only on$-\partial\mu j/\partial x$, but can be affected by any other force occurring in the system, such as the negative gradient in the chemical potential of some other species. A particular force, X_k, can likewise influence the flux of any species. Thus, the various fluxes become interdependent, or coupled, since they can respond to changes in any of the forces. Another premise in irreversible thermodynamics is that Jj is linearly dependent on the various forces. This means that in practice we will treat only those cases which are not too far from equilibrium. Even with such a simplification the algebra often becomes rather cumbersome, owing to the coupling of the various forces and fluxes. Using a linear combination of all the forces, we can represent the flux of species j by

$$J_j = \sum_k L_{jk} X_k = L_{j1} X_1 + L_{j2} X_2 + ... + L_j X_j + ... \quad ...(27)$$

where the summation, Σ_k, is over all forces (all X_k), and the L_{jk} are referred to as the Onsager coefficients, or the phenomenological coefficients, In this case for conductivity. The first subscript on these coefficients, e.g. j on Lj_k identifies the flux; the second subscript (i.e., k on Lj_k) designates the force. Each term, $Lj_k X_k$, is thus the partial flux of species j due to the particular force X_k. The individual Onsager coefficients in Equation 27 are therefore the proportionality factors indicating what contribution the force X_k makes to the flux of species *j*, *Jj* Equation 27 is sometimes referred to as the *phenomenological equation.*

Phenomenological equations are used to describe relations between observable quantities, without regard for explanations in terms of atoms or molecules. For instance, Ohm's law and Fick's laws are also phenomenological equations, likewise they assume linear relations between forces and fluxes. Are there any relations between the various phenomenological coefficients ? In fact, L_{jk} equals L_{kj}, which is known as the reciprocity relation. Such an equality of cross-coefficients was derived in 1931 by Onsager—awarded the Nobel Prize in chemistry in 1968—from statistical considerations which utilize the principle of "detailed balancing"—an urgent argument involving microscopic reversibility, i.e., for local equilibrium, any molecular process and its reverse will be taking place at the same average rate in that region. In discursive language, the Onsager reciprocity relation means that the proportionality coefficient relating the flux of species *k* caused by the force on species j equals the proportionality coefficient giving the flux of j due to the force on λ. (Strictly speaking, conjugate forces and fluxes must be used, as will be the case below. The fact that L_{jk} equals L_{kj} can be further appreciated by considering Newton's third law; action and reaction must be equal. For example, the frictional drag exerted by a moving solvent on the solute is equal to the drag exerted by the moving solute on the solvent. The pair-wise equality of cross-coefficients given by the Onsager reciprocity relation reduces the number of coefficients needed to describe the interdependence of forces and fluxes in irreversible thermodynamics and consequently leads to a simplification in solving the sets of simultaneous equations.

Water and Solute Flow

As a specific application of the principles just introduced, we will consider in some detail the important coupling of water and solute flow. The driving forces for the fluxes are the negative gradients in chemical potential, which we will assume are proportional to the differences in chemical potential across some barrier, here considered to be a membrane. In particular, we will represent— $\partial\mu j/\partial x$ by $\Delta\mu j/\Delta x$, which is $(\mu j^{\circ} - \mu j^{i})/\hat{e}x$ in the present case (For convenience, the thickness of the barrier, Δx, will be incorporated into the coefficient multiplying $\Delta\mu j$.) To help keep the algebra relatively simple, the development will be for a single nonelectrolyte. The fluxes are across a membrane permeable to both water

(w) and the single solute(s). From Equation 27, we can represent the fluxes as the following linear combination of the differences in chemical potential:

$$J_w = L_{ww}\Delta\mu_w + L_{ws}\Delta\mu_s \quad ...(28)$$

$$J_s = L_{sw}\Delta\mu_w + L_{ss}\Delta\mu_s \quad ...(29)$$

Equations 28 and 29 allow for the possibility that each of the fluxes may depend on the differences in both chemical potentials, $\Delta\mu_w$ and $\Delta\mu_s$. Four phenomenological coefficients are used in these two equations. By the Onsager reciprocity relation, L_{ws} equals L_{sw}. Thus, three different coefficients (L_{ww}, L_{ws}, and L_{ss}) are needed to describe the relationship of these two fluxes to the two driving forces. This is in constant to Equation Jj=ujcj (∂μj/∂x), where each flux depends on but one force; accordingly, only two coefficients would then be involved in describing J_w and J_s.

To obtain more convenient formulations for the fluxes of water and the solute, we again generally express $\Delta\mu_w$ and $\Delta\mu_s$ in terms of the differences in the osmotic and the hydrostatic pressures. $\Delta\pi$ and ΔP— it is usually easier to measure ΔP and $\Delta\pi$ that it is to determine $\Delta\mu_w$ and $\Delta\mu_s$. This is a straightforward matter for $\Delta\mu_w$; the only possible ambiguity is in deciding the algebraic sign. For the present case, $\Delta\mu_w$ is the chemical potential of water on the outside minus that on the inside, $\mu_w{}^o$—$\mu_w{}^i$. From equation ($\mu_w = \mu_w - \overline{V}_w\pi + \overline{V}_wP + m_wgh$), $\Delta\mu_w$ is given by

$$\Delta\mu_w = -\overline{V}_w\Delta\pi + \overline{V}_w\Delta P \quad ...(30)$$

where $\Delta\pi$ here equals $\pi^o - \pi^i$ and ΔP is $P^o - P^i$ ($\Delta h = 0$ across a membrane).

To express $\Delta\mu_s$ in terms of $\Delta\pi$ and ΔP, we must first consider the activity term, RT In a_s, in a little detail. We will begin with the differential RT d(In a_s), which equals RT da_s/a_s, or RT $d(\gamma_s c_s)/\gamma_s c_s$. When γ_s is constant, this latter quantity becomes simply RT dc_s/c_s. From Equation

$$\left(\pi_S = RT\sum_j cj\right),$$

RT dc_s equals $d\pi$ for a dilute solution of a single solute. Hence, RTd (In a_s) can be replaced by $d\pi/c_s$ as a useful approximation. In expressing the difference in chemical potential across a membrane we are interested in macroscopic changes, not in the infinitesimal changes given by differentials. To go from differentials to differences (Δ), RTd(In a_s) becomes RTΔ In a_s and $d\pi/c_s$ can be replaced by $\Delta\pi/c_s$, where c_s is essentially the mean concentration of solute s, in this case across the membrane. (Alternatively, we can simply define c_s as that concentration for which RTΔ In a_s exactly equals $\Delta\pi/c_s$.) Therefore, we can replace the term RTΔ In a_s in $\Delta\mu$, by the equivalent term, $\Delta\pi/c_s$. From Equation, μ_s equals μ_s + RT In $a_s + \overline{V}_s P$ for a neutral species, and the difference in chemical potential of the neutral solute across the membrane ($\Delta\mu_s$) becomes

$$\Delta\mu_S = \frac{1}{c_S}\Delta\pi + \overline{V}_S\Delta P \qquad ...(31)$$

The two expressions representing the driving forces, $\Delta\mu_w$ (Equation 30) and $\Delta\mu_s$ (Equation 31), are expressed as functions of the same two pressure differences, $\Delta\pi$ and ΔP, the latter being rather convenient from the point of view of experiment.

Volume flow, L_P, and σ

Now that we have appropriately expressed the chemical potential differences of water and the solute, we will direct our attention to the fluxes. In our usual units, the fluxes J_w and J_s are the moles of water and solute, respectively, moving across one cm^2 of membrane surface in a second. A quantity of considerable interest in plant and cell physiology is the volume flow, Jv, which is the rate of movement of the total volume of both water and solute across unit area of the membrane. Jv has the units of volume per unit area per unit time, e.g., cm^3/cm^2 –sec), which is cm/sec. The molar flux of some species j(Jj) in moles/cm^2-sec times the volume occupied by each mole of it ($\overline{V}j$) in cm^3/mole gives the volume flow for that component (Jvj)/in cm/sec. For solute and water both moving across the membrane, the total volume flow Jv is the volume flow of water plus that of solute, which in the case of a single solute can be represented as follows:

$$Jv = V_w J_w + V_s J_s \qquad ...(32)$$

It is generally simpler and more convenient to measure the total volume flux (such as that given in Equation 32 than one of the component volume fluxes ($Jvj = \overline{V}jJj$). For instance, we can often determine the volumes of cells or organelles under different conditions and relate any changes in volume to Jv. The volume flux of water, Jv_w is simply $\overline{V}_w J_w$. However straightforward it may be, the algebraic substitution necessary to incorporate the various forces and fluxes into the volume flow given by Equation 32 leads to a rather cumbersome expression. Nevertheless, it provides—albeit in symbols—insight into the origin of the reflection coefficient, an important quantity that we will consider in detail below. When the fluxes of water and solute given by Equations 28 and 29, respectively, substituted into the expression for the volume flow (Equation 32), we obtain the follow in relation:

$$Jv = (V_w L_{ww} + V_s L_{sw})\,\Delta\mu w + (V_w L_{ws} + V_s L_{ss})\,\Delta\mu \quad ...(33)$$

As we mentioned above, L_{sw} equals L_{ws}, so that only the symbol L_{ws} need be retained here. Next, we will incorporate the chemical potential differences across the membrane for water and solute, as given by the Equation 30 and 31 respectively, into Equation 33. Upon rearrangement, $J\nu$ is then as follows:

$$Jv = (\overline{V}_w^2 L_{ww} + 2\overline{V}_s \overline{V}_w L_{ws} + \overline{V}_s^2 L_{ss})\Delta P$$
$$-(\overline{V}_w^2 L_{ww} - V_w \frac{1}{c_s} L_{ws} - \overline{V}_s \frac{1}{c_s} L_{ss}$$
$$+ V_s V_w L_{ws})\Delta\pi \quad ...(34)$$

Equation 34 gives the dependence of the total volume flow on the hydrostatic and osmotic pressure differences across some membrane permeable to both water and the solute. The quantities multiplying ΔP and $\Delta\pi$ in Equation 34 are rather cumbersome, so we will find it convenient to define some new parameters. The factor multiplying the difference in hydrostatic pressure, ΔP, is called the *hydraulic conductivity coefficient*, L_P. From Equation 34, this hydraulic conductivity coefficient has the following value.

$$L_P = \overline{V}_w^2 L_{ww} + 2\overline{V}_s \overline{V}_{ws} + \overline{V}_s^2 L_{ss} \quad ...(35)$$

Such a coefficient describes the membrane conductivity, since the larger L_P the greater is Jv for a given hydrostatic pressure difference cross the membrane. The other unwieldly factor appearing in Equation 34, viz., the parenthesis multiplying $\Delta\pi$, is incorporated into the reflection coefficient, σ. It is advantageous to define σ as the ratio of the coefficient of $\Delta\pi$ divided by the coefficient of ΔP in the expression for the volume flow (Equation 34). Using the definition of L_P (Equation 35), σ is therefore given by

$$\sigma = \frac{\overline{V}_w^2 L_{ww} - \overline{V}_w \frac{1}{c_s} L_{ws} - \overline{V}_s \frac{1}{c_s} L_{ss} + \overline{V}_s \overline{V}_w L_{ws}}{L_P} \quad ...(36)$$

The reflection coefficient defined by Equation 36 is perhaps the most important parameter from irreversible thermodynamics relevant to the fluxes generally considered in plant physiology. It was introduced by Staverman in 1948 and specifically applied to biological situations by Kedem and Katchalsky in 1958, numerous applications of irreversible thermodynamics, in general, and reflection coefficients, in particular, have been made in biology. We will shortly consider specific values of σ for solutes entering chloroplasts. We will also show that the maximum value of the reflection coefficient occurs when the solute is unable to penetrate the membrane-the molecules are then all *reflected* from the barrier.

In the absence of a hydrostatic pressure difference across a membrane, the volume flow (Jv) equals—$L_p\sigma\Delta\pi$ by Equation 34. Thus, the magnitude of the dimensionless parameter s determines the volume flow expected in response to a difference in osmotic pressure across a membrane—it is this use which is most pertinent in plant physiology. After further considering the expression for volume flow, we will discuss the properties of reflection coefficients (σ is derived in a more conventional). Using L_p and s as defined by Equation 35 and 36 respectively, we can rewrite the expression for the total volume flow (Equation 34 in the following convenient form:

$$Jv = L_p(\Delta P - \sigma\Delta\pi) \qquad ...(37)$$

Equation 37 indicates that the osmotic contribution to the volume flow across the membrane is changed by the factor σ, which is less than one when the solute can across the membrane. Many different solutes can penetrate a biological membrane under usual conditions. Each such species *j* can be characterized by its own reflection coefficient, σj, for that particular membrane. The volume flow given by Equation 37 can then be generalized as

$$J_V = L_P(\Delta P - \sum_j \sigma j \Delta\pi j) \qquad ...(38)$$

where Δπj is the osmotic pressure difference across the membrane for species *j* (e.g., Δπj = RTΔcj. Although interactions with water are still taken into account, the generalization represented by Equation 38 introduces the assumption that the solutes do not interact with each other as they penetrate the membrane. Moreover, J*j* in Equation 38 refers only to the movement of neutral species—otherwise we would also need a current equation to describe the flow of charge. Nevertheless, Equation 38 is a useful approximation of the actual situation, describing the multicomponent solutions encountered by cells, and we with use it as the starting point for our general consideration of solute movement across membranes. Before discussing such movement, however, let us consider the range of value of reflection coefficients.

Values of Reflection Coefficients

A reflection coefficient characterizes some particular solute interacting with a specific membrane. In addition, σj depends on the solvent on either side of the membrane, as the subscripts w for water in Equation 36 clearly indicate—is the only such solvent we will consider here. Two extreme conditions can describe the passage of solutes: impermeability, which leads to the maximum value of unity for the reflection coefficient, and nonselectivity, where σj is zero.

A reflection coefficient of zero may describe the movement of a solute across a very coarse barrier (one with large pores) which does not distinguish or select between solute and solvent molecules, or it may refer to the passage through a membrane of a molecule very similar in size and structure to water itself. Impermeability describes the limiting case where water can cross some membrane, but the solute cannot. When the solute does not penetrate the membrane at all. J_s must remain zero must remain zero any and all values of $\Delta\mu_w$ and $\Delta\mu_s$. Since J_s equals $L_{sw}\Delta\mu_w + L_{ss}\Delta\mu_s$ (Equation 29), the only way in which J_s will always be zero is for both L_{ss} and L_{ws} to be zero.

Therefore, when the membrane is impermeable to the particular solute under consideration, both L_{ss} are zero. Putting these zero values for L_{ss} and Lw_s into Equation 35 and 36, we obtain the following two relations: $Lp=\overline{V}_w^{\,2} L_{ww}$ (Equation 35) and $\sigma=V_w^{\,2} L_{ww}/ L_P$ (Equation 35). Consequently, σ then equals $\overline{V}_w^{\,2}L_{ww}/\overline{V}_w^{\,2} L_{ww}$, which is one. A reflection coefficient of unity thus signifies that the solute is not crossing the membrane, i.e., all of the solute is "reflected" from the membrane, and the reflection coefficient has its maximum value. When a solute can penetrate, L_{ss} takes on a positive value, i.e., the force on solute *s* causes a flux of it across the membrane. The $L_{ss}\overline{V}_s/C_s$ term then causes the numerator of the expression for σ (Equation 36) to decrease as Lss increases, while the denominator (L_P, see Equation 35) tends to increase as the solute's ability to penetrate goes up.

The reflection coefficient thus becomes less than one when the solute can cross the membrane. This last comment emphasizes the fact that the general case of solute permeability. Moreover, the two terms containing L_{ws} in the numerator of the expression for the reflection coefficient (Equation 36) take into account the further possibility that solute and solvent may interact when crossing the membrane, i.e., σ can also be affected by coupling or independence of forces and fluxes. When a membrane or barrier is nonselective, both water and the solute move across it at the same velocity, i.e., $v_s=v_w$. The solution is then the same on both sides of the barrier, so that $\Delta\pi$ across the barrier is zero, which means that $\Delta\mu w$ equals $\overline{V}w\Delta P$ and $\Delta\mu_s$ is $\overline{V}_s\Delta P$. (Admittedly, the idea of a solution containing a single solute is not realistic from a biological point of view, but it is convenient for illustrating the minimum value for σ).

As we discussed above vj is simply **Jj/c*j*. Using these various conditions as well as Equations 28 and 29, we obtain the following string of equalities:** $v_w = J_w/c_w = L_{ww}\Delta\mu_w/c_w + L_{ws}\Delta\mu_s/C_w = (L_{ww}\overline{V}_w/c_w + L_{ws}\overline{V}s/c_w)\Delta P = v_s = J_s/c_s = L_{ws}\ \Delta\mu_w/c_s + L_{ss}\Delta\mu s/c_s = (Lw_s\ \overline{V}_w/c_s + L_{ss}\ \overline{V}_s/cs)$ DP. Next, we will let the number of moles of water per unit volume (c_w) times the volume per mole of water ($\overline{V}_w$) equal unity—$c_w\overline{V}_w \cong 1$ is another way of stating the dilute solution approximation. Also, we will identify c_s with c_s. Putting these last two relations into the factors multiplying ΔP in the above string of equalities yields $L_{ww}\overline{V}_w{}^2 + L_{ws}\overline{V}_s\overline{V}_w = L_{ws}\overline{V}_w/c_s + Ls_s\overline{V}_s/c_s$. Finally, we can use this desired relation to evaluate the numerator of Equation 36. Accordingly σ equals $0/L_p$, which is zero, i.e., for the case of non selectivity ($v_s = v_w$), the reflection coefficient of the solute must be zero for that barrier. Another relatively simple condition by which σ equals zero can be readily understood from Equation 37 [$J_v = L_P(\Delta P - \sigma\Delta\pi)$].

Let us consider the rather realistic situation where ΔP is zero across a membrane. The volume flow (Jv) is then simply $- L_P\sigma\Delta\pi$, by Equation 37. For a solute having a reflection coefficient equal to zero for that particular membrane, the volume flow would be zero. By Equation 32 ($J_v = V_wJ_w + \overline{V}_sJ_s$), a zero J_v implies that $\overline{V}_wJ_w$ equals—$\overline{V}_sJ_s$. In Words, the volume flow of water must be equal and opposite to the volume flow of the solute in order to get no net volume flow. It follows that when ΔP is zero but Δπ is non zero, the absence of a net volume flow J, across a membrane permeable to both water and the single solute indicates that the reflection coefficient for that solute is zero. This condition of σ equaling zero occurs when the volume of water flowing toward the side with the higher π is balanced by an equal volume of solute diffusing across the membrane in the opposite direction toward the side where the solute is less concentrated (lower π). Such a situation of zero volume flow anticipates the concept of a "stationary state," introduced in the next section.

Solutes interacting with biological membranes exhibit properties ranging from impermeability (σj = 1) all the way to nonselectivity (σj = 0). Substances which are retained in, or excluded from plant cells have reflection coefficients rather close to unity for the cellular membranes. For instance, the σ*j*s for sucrose and amino acids are usually near 1.0 for plant cells. Methanol and ethanol enter cells very readily and nave reflection coefficients of approximately 0.3 for some *Chara* and *Nitella* internodal cells. On the other hand, σj can essentially equal zcro for solutes crossing porous barriers, such as those presented by cells walls, or for molecules that are penetrating very readily across membranes, such as D_2O. As is the case for a permeability coefficient, the reflection coefficient of some species is the same for traversal in either direction across a membrane. For many small neutral solutes nor interacting with carriers in the membrane, the reflection coefficients are correlated with the partition coefficients. For example, when Kj for nonelectrolytes is less than about 10^{-8}, σj is generally close to one. Thus compounds which do not readily enter the lipid phase of membrane (low Kj) also do not cross the membrane easily (σj near 1).

When the partition coefficient is one or greater, the solutes can enter the membrane in appreciable amounts, and σj is generally close to zero. Considering the intermediate case, the reflection coefficient of some species j can be near 0.5 for a small nonelectrolyte having a Kj of about 0.1 for the membrane lipids, although individual molecules differ depending on their molecular weight, branching, and atomic composition. Neglecting frictional effects with other solutes, we see that the intermolecular interactions affecting the partition coefficient are similar to those governing the value of the reflection coefficient, and there is somewhat of a correlation therefore between the permeability coefficient of a solute ($Pj = D_jK_j/\Delta x$) and its reflection coefficient for the same membrane. In fact, there is even a correlation between the reflection coefficients of a series of nonelectrolytes determined with animal membranes and the permeability coefficients of the same substances measured for plant membranes. Thus, as the permeability coefficient goes from very small to very large values, the reflection coefficient decreases from unity (describing relative impermeability) to zero (for the opposite

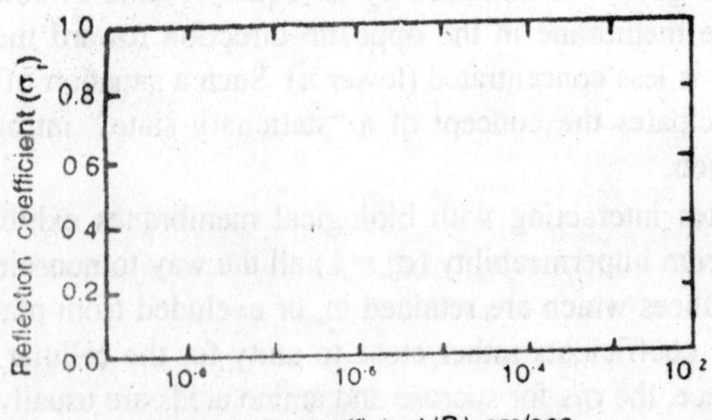

Fig. 4.6. Correlation between the reflection coefficients for a series of nonelectrolytes (determined using rabbit gallbladder epithelium) and the permeability coefficients for the same compounds (measured by R. Collander using Nitella mucronata).

extreme of nonselectivity). In the next section we will specifically consider some of the consequences of reflection coefficients differing from unity for solutes crossing cellular and organelle membranes.

Solute Movement Across Membranes

We can profitably reexamine certain aspects of the movement of solutes into and out of cells and organelles by using the more general equations developed from irreversible thermodynamics. One particularly important situation amenable to rather uncomplicated analysis is when the total volume flow J_v is zero, an example of a *stationary state*. This stationary state, in which the volume of the cell or organelle does not change over the time period of interest, can be brought about by having the net volume flow of water in one direction across the membrane equal the net volume flow of solutes in the opposite direction. A stationary state is therefore not be same as a steady state or an equilibrium condition for one cellular organelle; it represents a situation occurring only at a particular time or under some special experimental arrangements in this regard we should point out that μ_j can depend on both position and time for a stationary state but only on position form steady state, and on neither position nor into at equilibrium. When water is moving into a cell, $\Delta\mu_w$ is non-zero (we are not at equilibrium) and, in general μ_w will be increasing with time.

If over some time interval the volume flow solutes out of the cell equals this J_v, in the cell will be in a *stationary state*, where its volume is not changing. One restriction here to cases of zero volume flow considerably simplifies the algebra and emphasizes the role played by reflection coefficients. Moreover, a stationary state of no volume change often characterizes the experimental situation under which the Boyle-Van't Hoff relation or the expression describing incipient plasmolysis. What is the relationship between the internal and the external Ps and πs in the case of a stationary state ?In order to answer this we must first express the consequences of a stationary state in symbols. Our point of departure is Equation 38

$$J_V = L_P \Delta P - \sum_j \sigma_j \tau_j,$$

where the stationary state condition of zero net volume flow ($J_v = 0$) leads to the following equalities

$$0 = \Delta P - \sum_j \sigma_j \sigma \tau_j = P^o - \sum \sigma (\tau_j^o - \tau_j^i)$$

Here, the osmotic pressures are for the effect of solutes on water activity, while in the general case both osmotic contributions from solutes (τ_s) atmospheric pressures resulting from the presence of interfaces (τ) might occur. Volume made when the external solution is at atmospheric pressure ($P^o = 0$) and when there are no external interfaces ($\tau = 0$). The stationary state condition of $J\nu$ equaling zero then leads to

$$\sum \sigma_j \pi_j^o = \sum_j \sigma_j \pi_j^i + \tau^i - P^i$$

where the possibilities of interfacial interaction and hydrostatic pressures within the cell or organelle are explicitly recognized by the inclusion of τ^i and P^i. Equation 39 describes the stationary state of zero volume flow. It applies when the external solutions is at atmospheric pressure (and τ^o = 0) and when the solutes are capable of crossing the barrier, as occurs for molecules interacting with real, not idealized, biological membranes.

The Influence of Reflection Coefficients on Incipient Plasmolysis

We used classical thermodynamics to derive the condition for incipient plasmolysis ($\pi^o_{plasmolysis} = \pi^i$), which occurs when the internal pressure P^i inside a plant cell just becomes zero. This derivation was made assuming equilibrium of water, i.e. equal water potential, across a membrane impermeable to solutes. The assumptions of water equilibrium and impermeability are often not valid. Can we rectify the situation using an approach based on irreversible thermodynamics? Measurements of incipient plasmolysis can be made when there is zero volume flow (Jv = 0) and for simple external solution ($\tau^o = 0$) at atmospheric pressure (P^o = 0). In this case, Equation 39

$$\left(\sum_j \sigma_j \pi_j^o = \sum_j \sigma_j \pi_j^i + \tau^i - P^i\right)$$

would be the appropriate underlying expression from irreversible thermodynamics, instead of the less realistic condition of water equilibrium previously used. Using this stationary state conditions, we obtain the following expression describing incipient plasmolysis ($P^i = 0$) when the solutes can cross the cell membrane:

$$\sigma^o \pi^o{}_{plasmolysis} = \sum_j \sigma_j \pi_j^i + \tau^i \quad ...(40)$$

where

$$\sum_j \sigma_j \pi_j$$

has been replaced by $\sigma^o \pi^o$. In other words, it is often convenient to characterize the external solutes by an average reflection coefficient, σ^o, where

$$\sum_j \sigma_j \pi_j^o$$

equals $\sigma^o \pi^o$ and π^o is the total external osmotic pressure i.e.,

$$\pi^o = \sum_j \pi_j^o$$

Since the value of σ^o δεπενδσ on the particular external solutes present, Equation 40 indicates that the external osmotic pressure (π^o) at incipient plasmolysis varies from solute to solute placed in the solution surrounding the plant cells. Let us suppose that solute i cannot penetrate the membrane, so that σi equals 1 a situation often true for sucrose. Another solute, j, can penetrate into the cells (σj < 1), as is the case for many small nonelectrolytes. If we are at the point of incipient plasmolysis for each of these two solutes—taken in turn as the sole species in the external solution—$\sigma_i \pi_i{}^o{}_{\text{plasmolysis}}$ equals $\sigma_i \pi_i{}^o{}_{\text{plasmolysis}}$ by Equation 40. But solute i is unable to penetrate membrane ($\tau_i = 1$). Hence we obtain the following relationships:

$$\sigma_j = \frac{\tau^o_{\text{i plasmolysis}}}{\pi^o_{\text{j plasmolysis}}} = \frac{\text{effective osmotic pressure of species j}}{\text{actual osmotic pressure of species j}} \quad ...(41)$$

Equation 41 suggests a rather straightforward way of describing σj. Since (by supposition) species *j* can cross the cell membrane, σj must be less than one. Therefore, by Equation 41, $\pi j^o{}_{\text{plasmolysis}}$ is greater than $\pi_i{}^o{}_{\text{plasmolysis}}$ where the latter refers to the osmotic pressure of the impermeable solute at the point of incipient plasmolysis. In words, a higher external osmotic pressure is needed to cause plasmolysis if that solute is able to enter the plant cell. The actual osmotic pressure indicated in Equation 41 is defined either by $\pi = -(RT/V_w) \ln a_w$, where

$$\pi = \sum_j {}^o\pi_j, \text{ or by } \pi_j = RT\gamma_j c_j,$$

when activity coefficient are included. When the membrane is impermeable to the solute, σj equals one, and the apparent (effective) osmotic pressure of species j equals the actual σj°. Effective osmotic pressure takes into consideration the fact that most solutes can move across biological membranes (σj < 1), hence the πj effective in leading to a net volume flow is reduced from its actual value—cf Equation 38,

$$J_V = L_P(\Delta P - \sum_j \sigma_j \Delta_j)$$

In summary, the reflection coefficient of species j indicates how effectively the osmotic pressure of that solute can be exerted across a particular membrane or other barrier. We can also use the condition of incipient plasmolysis described by Equation 41 to evaluate specific reflection coefficients.

It is difficult in experiments to replace one external solution by another with no changes in the tissue taking place, or with none of the previous solution adhering to the cell. Also, although it is easy in principle and involves only the use of a light microscope, determination of when the plasmalemma just begins to pull away from the cell wall is a rather subjective judgement. Nevertheless, the use of Equation 41 provides a simple way of considering individual reflection coefficients for various solutes entering plant cells. Since osmotic pressures play a key role in plant physiology, σ js are important parameters for quantitatively describing the solute and water relations of plants.

In particular, the incorporatic of reflection coefficient allows the role of osmotic pressure to be precisely stated. In Poiseuille" law. Jv = – (r² / 8n) θP/∂x which can adequately describe movement in the xylem and possibly in the phloem, then flow is driven by the hydrostatic pressure gradient. An alternative view of the same situation is that σ equals zero for the solutes. In that case, osmotic pressures would have no direct effect on the movement described by Poiseuille's law. The real interest in reflection coefficient comes when σ*j* is not at one of its two extremes of zero and unity. For such cases—intermediate between non-selectivity and impermeability—the volume flow does not depend on the full osmotic pressure difference across the barrier, but would be invalid to completely ignore the osmotic contribution of species j. To illustrate some of the points we have been making concerning incipient plasmolysis and reflection coefficients.

The cell in the upper left of the diagram is at the point of incipient plasmolysis ($P^i = 0$) for a nonpenetrating solute (σ° = 1.0) in the external solution (π° = A, where A is a constant) i.e., the plamalemma is just beginning to pull away from the cell wall. Equation 40 indicates that σjπj would then also equal A. If we place the cell in a second solution containing a penetrating solute (σ° <1), Equation 41 indicates that the external solution must have a higher osmotic pressure in order for the cell to remain at the point of incipient plasmolysis. For instance, for a second solution with a σj of 0.5, the external osmotic pressure at the point of incipient plasmolysis would be (2A). Thus, when the external solute can penetrate, π° is not as effective in balancing the internal osmotic pressure c in leading to a flow of water. On the other hand if π° were 2A for an impermeable external solute, extensive plasmolysis of the same cell would occur. Since $\sigma^\circ\pi^\circ$ is 2A in this case, Equation 39 ($\sigma^\circ\pi^\circ = \sigma^i\pi^i - P^i$, when $\pi^I = 0$) indicates that $\sigma^i\pi^i$ must also be 2A which means that essentially half of the internal water has left the cell.

Finally if the reflective coefficient were 0.5 and the external osmotic pressure were A, $\sigma^\circ\pi^\circ$ would be ½A, and we would not be at the point of incipient plasmolysis. In fact, the cell would be under turgor with an internal hydrostatic pressure equal to ½A, at least until the concentration of the penetrating solute began to build up inside. We must therefore take into account the reflection coefficients of external (and internal) solutes in order to describe conditions at the point of incipient plasmolysis and, by extension, predict the direction and the magnitude of volume fluxes across membranes.

Extension of the Boyle-Van't Hoff Relation

According to Boyle-Van't Hoff relation the water potential was the state on both sides of the cellular or organelle membrane under consideration. Net only were equilibrium conditions imposed on water, but we implicitly assumed that the membrane was impermeable to the solutes. However, zero total volume flow (Jv = 0) is a better description of the experimental situation where the Boyl-Van't Hoff relation is actually applied. This condition of no volume change during the measurement is another example of a stationary state, so the Boyle-Van't Hoff relation with be reexamined from the point of view of irreversible thermodynamics. In this way we can remove two of the previous restriction, equilibrium for water and impermeability of solutes. When molecules can cross the membranes bounding cells or organelles, the reflection coefficients for both internal and external solutes would be included in the Boyle-Van't Hoff relation. Since σ° is less than one when the external solutes can penetrate, the effect of the external osmotic pressure on J*v* is reduced. Likewise, the reflection coefficients for solutes within the cell or organelle lessen the contribution of the internal osmotic pressure of each solute. Replacing

$$\sum_j \sigma_j \pi_j^o$$

by $\sigma^\circ\pi^\circ$ and πj^i by RT $\gamma_j^i n_j^i$/(Vwηw^i). Equation 39 leads to the following Boyle-Van't Hoff relation for the stationary state condition (Jv=0 in Eq. 38)

$$\pi^\circ = RT\frac{\sum_j \sigma_j \gamma j^i nj^i}{\sigma^\circ V_w nw^i} + \frac{\tau^i - P^i}{\sigma^\circ} \quad ...(42)$$

The reflection coefficients of the membrane for both internal and external solutes enter into this extension of the expression relating volume and external osmotic pressure. The quantity V – b in the conventional Boyle-Van't Hoff relation,

$$\pi^\circ(V-b) = RT\sum_j \varphi jnj$$

can be identified with $V_w n_w^i$. If we see Equation 42, it is evident that the osmotic coefficient of species j, φ_j can be equated to $\sigma j\gamma j^i/\sigma^\circ$ as an explicit recognition of the activity coefficient of species j and the permeation properties of solutes, both internal and external. In fact, failure to recognize the effect of reflection coefficients on φj has led to misunderstandings on osmotic responses. The volume of pea chloroplasts responds linearly to $1/\pi^\circ$ indicating that $\tau^i - P^i$ in such plastids may be negligible compared with the external osmotic pressures used.

To analyze experimental observations, it is convenient to replace $\sigma^\circ\pi^\circ$ by $\sigma_x\pi_x{}^\circ + \alpha$, where $\pi_x{}^\circ$ is the contribution to the external osmotic pressure of solute x whose reflection coefficient (σ_x) is being considered, and α is the sum of $\alpha j\pi j^\circ$ for all the other external solutes. We can represent

$$RT\sum_j \sigma_j \gamma j^i nj^i$$

by β, and replace $V_w n_w{}^i$ by V–b. Making these substitutions into Equation 42, we obtain the following relatively simple form for the Boyle-Van't Hoff relation in the case penetrating solutes:

$$\sigma_x\pi_x^o + \alpha = \frac{\beta}{V-b} \quad ...(43)$$

If we vary $\pi_x{}^\circ$ and measure V, we can then use Equation 43 to obtain the reflection coefficient for various nonelectrolytes in the external solution.

Reflection Coefficients for Chloroplasts

When the refinements introduced by reflection coefficients are taken into account, we can use osmotic responses of cells and organelles to quantitatively describe the permeability properties of their membranes. As a specific application of Equation 43, the progressive addition of hydroxymethyl groups in a series of polyhydroxy alcohols causes the reflection coefficients to steadily increase from 0.00 to 1.00 for pea chloroplasts. In this regard, the lipid : water solubility ratio decreases in going from methanol to ethylene glycol to glycerol to erythritol to adonitol, i.e., the partition coefficient K_x decreases. Since the permeability coefficient P_x equals $D_xK_x/\Delta x$ we expect a similar decrease in P_x as hydroxymethyl groups are added. Figure clearly shows that as the permeability coefficient decreases the reflection coefficient generally increases. Consequently, the increase in σ_x of alcohols as—CHOH groups are added can be interpreted as simply a lowering of K_x. (As we go from methanol to adonitol, D_τ also decreases, perhaps by a factor of 3, while K_x decreases about a thousandfold so that changes in K_x are the predominant influence on P_x and σ_x in this case).

The reflection coefficients of six-carbon polyhydroxy alcohols, such as sorbitol and mannitol, are essentially unity for pea chloroplasts. This indication of relative impermeability suggests that these compounds would serve as suitable osmotica in which to suspend chloroplasts, as is indeed the case. We may reasonably expect partition coefficients or the chloroplast membranes to be lower for amino acids than for adonitol. Since $\sigma_{adonitol}$ is 1.00 for pea chloroplasts, we would therefore anticipate that the reflection coefficients of amino acids would be near unity. However, this is not the case, as five of the amino acids have a σ_x of 0.4 or less for pea chloroplasts. L-isoleucine ($\sigma x = 0.33$), L-valine (0.53), and L-leucine (0.56) have somewhat higher reflection coefficients, but these amino acids all have a branch methyl group on the side chain which apparently hinders their passage across the chloroplast membranes. In contrast, the σ_xs of these same eight amino acids are in all cases near unity for the cellular membranes which have been studied.

The fact that the reflection coefficients of amino acids are low for chloroplasts and high for cells has important physiological implications. Amino acids needed for protein synthesis in chloroplasts can then readily diffuse in from the cytoplasm. Also, amino acids produced as photosynthetic products can move out into the cytoplasm without the utilization of any energy for active transport. On the other hand, if amino acids could readily diffuse across cellular membranes, they would rapidly be lost from the cell. Since the reflection coefficients of amino acids for pea chloroplasts are much less than predictions based on the partition coefficients, the permeation of amino acids into the chloroplasts may not be governed by the partition of these compounds between membrane lipids and water.

Evidently, carriers passively shuttle the amino acids into and out of chloroplasts and thereby lower the effective reflection coefficients. When the concentration of an amino acid is raised above 100 mM in the external solution bathing the chloroplasts, the reflection coefficient increase to values near unity. This is interpreted as a saturation of the carriers used for facilitated diffusion, and can be analyzed using a Michaelis-Menten type of formalism (Equation 26). If an aliphatic amino acid—L-alamine, L-isoleucine, L-valine, and L-leucine-is placed in the external solution with glycine, the chloroplast osmotic responses indicate competition for the same carrier. A second carrier for L-serine, L-threonine, and L-methionine is also demonstrated. In short, three criteria for the use of carriers in facilitated diffusion, viz., specificity, saturation, and competition, can all be demonstrated using the osmotic responses of chloroplasts as interpreted by the modified from of the Boyle-Van't Hoff relation. Hence, the rather esoteric concepts of irreversible thermodynamics can be applied in a relatively simple manner to gain insights into the physiological attributes of membranes.

5

PHLOEM TRANSLOCATION

Survival on land posed a new set of problems for plants newly emerged from an aquatic environment—the need to acquire and retain water being primary among them. In response to these environmental pressure, plants evolved roots (which anchor the plant and absorb water and nutrients) and leaves (which absorb light and exchange gases). As plants increased in size, the roots and leaves became increasingly separated from each other in space. Thus, systems evolved for long-distance transport—that is, translocation—that allowed the shoot and the root to efficiently exchange the products of absorption and assimilation. The *xylem* is the tissue that transports water and minerals from the root system to the aerial portions of the plant. The *phloem*, which is the subject of this chapter, is the tissue that translocates the products of photosynthesis from mature leaves to areas of growth and storage, including the roots. As we shall see, the phloem also serves to redistribute water and various compounds throughout the plant body. These compounds, some of which initially arrive in the mature leaves via the xylem, can be either transferred out of the leaves without modification or metabolized before redistribution.

In the discussion that follows, the emphasis will be on phloem translocation in angiosperms, since most of the research has been conducted on that group of species. Gymnosperms will be briefly compared to the angiosperms in terms of the anatomy of their conducting cells and possible differences in their mechanism of translocation.

PATHWAYS OF TRANSLOCATION

As mentioned above, two long-distance transport pathways, the phloem and the xylem, extend throughout the plant body. The phloem is

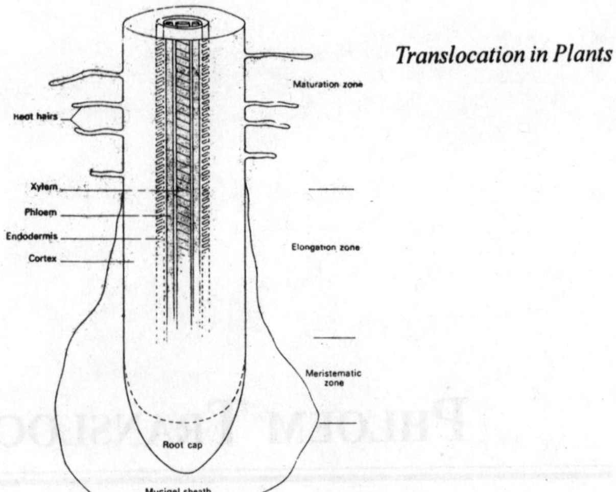

Fig. 5.1. Transverse section vascular bundle.

generally found on the outer side of both primary and secondary vascular tissues; in plants with secondary growth the phloem constitutes the inner bark. The cells of the phloem that conduct sugars and other organic materials throughout the plant are called *sieve elements*. "Sieve element" is a comprehensive term that includes both the highly differentiated *sieve-tube members* typical of the angiosperms and the relatively unspecialized *sieve cells* of gymnosperms. In addition to sieve elements, the phloem tissue contains companion cells, parenchyma cells, and, in some cases, fibers, sclereids, and latex-containing cells (laticifers). However, only the sieve elements are directly involved in translocation. The small veins of leaves and the primary vascular bundles of stems are often surrounded by a *bundle sheath*, which consists of one or more layers of compactly arranged cells. In the vascular tissue of leaves, the bundle sheath surrounds the small veins all the way to their ends, isolating the veins from the intercellular spaces of the leaf. In some leaves, cells similar to the bundle sheath cells extend to the upper and lower epidermal layers and are thought to aid in water conduction throughout the leaf.

We will begin our discussion with the experimental evidence demonstrating that the sieve elements are the conducting cells in the phloem. This will be followed by an examination of the structure and physiology of these unusual plant cells.

Experiments with Radioactive Labels Show That Sugar is Translocated in Phloem Sieve Elements

In classical experiments on the translocation of organic solutes performed by Malpighi in 1686, the bark of a tree was removed in a ring

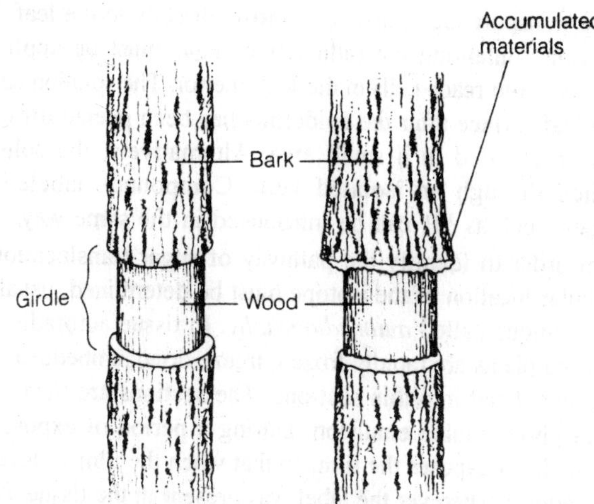

Fig. 5.2. Tree trunk immediately after girdling (left) and after a longer period of time (right). Girdling is the removal of the bark of a tree in a ring around the trunk. Meterials translocated from the leaves have accumulated in the region above the girdle and caused it to swell.

around the trunk. In 1928 Mason and Maskell observed that this treatment, called *girdling*, has no immediate effect on transpiration, since water movement occurs in the xylem, interior to the bark. However, sugar transport in the trunk is blocked at the site where the bark has been removed. Sugars accumulate above the girdle, that is, on the side toward the leaves, and are depleted below the treated region. Eventually the bark below the girdle dies, while the bark above swells and remains healthy. Mason and Maskell concluded that sugar transport occurs in the bark of the tree and, further, that the sieve elements are the cellular channels of sugar transport. The latter conclusion was based on an observed high correlation between leaf and bark sucrose contents and on the high sucrose concentration calculated to be present in the sieve elements.

More sophisticated experiments on phloem translocation became possible in the 1940s when radioactive isotopes became available for scientific research. Labeled organic compounds can be introduced into the plant in a number of ways. For example, carbon dioxide can be labeled with ^{14}C or ^{11}C and supplied to an intact mature leaf in a small sealed chamber. The labeled CO_2 is incorporated into organic compounds via the CO_2-fixing reactions of photosynthesis. Some of the sugar-phosphates formed in photosynthesis undergo further reactions to form the translocated sugars, such as sucrose and stachyose. In another type of experiment, the photosynthetic reactions are bypassed by supplying labeled transport sugar, usually sucrose, directly to the leaf. In this case, a solution containing the radioactive sugar must be applied in a way that allows it to reach cells in the leaf interior. The solution can be applied to the leaf surface after the epidermis has been peeled off or the cuticle has been abraded in a small area. Alternatively, the solution can be supplied through an isolated vein. Compounds labeled with other isotopes, such as ^{32}P, can be introduced in the same way.

In order to identify the pathway of sugar translocation, the tissue or cellular location of the isotope must be determined, usually by means of a technique called *autoradiography*. In tissue autoradiography, parts of labeled plants are rapidly frozen, freeze-dried, embedded in paraffin or resin, and sliced into thin sections. The sections are then coated with a film of photographic emulsion. During a period of exposure, radiation from the label exposes the film, so that when the film is developed, silver grains appear wherever the label was present in the tissue. A comparison of the tissue section and the pattern of silver grains reveals the location of the label in the tissue. In terms of the transport pathway for sugars, the label initially appears in the sieve elements of the phloem, confirming the results of the earlier experiments.

Mature Sieve Elements are Living Cells that are Highly Specialized for Translocation

As we will see later, detailed knowledge of the ultrastructure of sieve elements is critical to any discussion of the mechanism of phloem translocation. Mature sieve elements are indeed unique among living plant cells. They lack many structures normally found in living cells, including the undifferentiated cells from which mature sieve elements are formed. For example, sieve elements lose their nucleus and tonoplast (vacuolar membrane) during development. Microfilaments, microtubules, Golgi bodies, and ribosomes are also absent from the mature cells. In addition to the plasma membrane, organelles that are retained include somewhat modified mitochondria, plastids, and smooth endoplasmic reticulum. The walls are nonlignified, though they are secondary thickened in some cases. Thus, the sieve elements are unlike the tracheary elements of the xylem, which are dead at maturity, lack a plasma membrane, and have lignified secondary walls. As we shall see, this difference is critical to the mechanism of translocation in the phloem.

The Most Characteristic Feature of the Sieve Elements is the Presence of the Sieve Areas

Sieve elements are characterized by *sieve areas*, portions of the cell wall where pores interconnect the conducting cells. The sieve-area pores

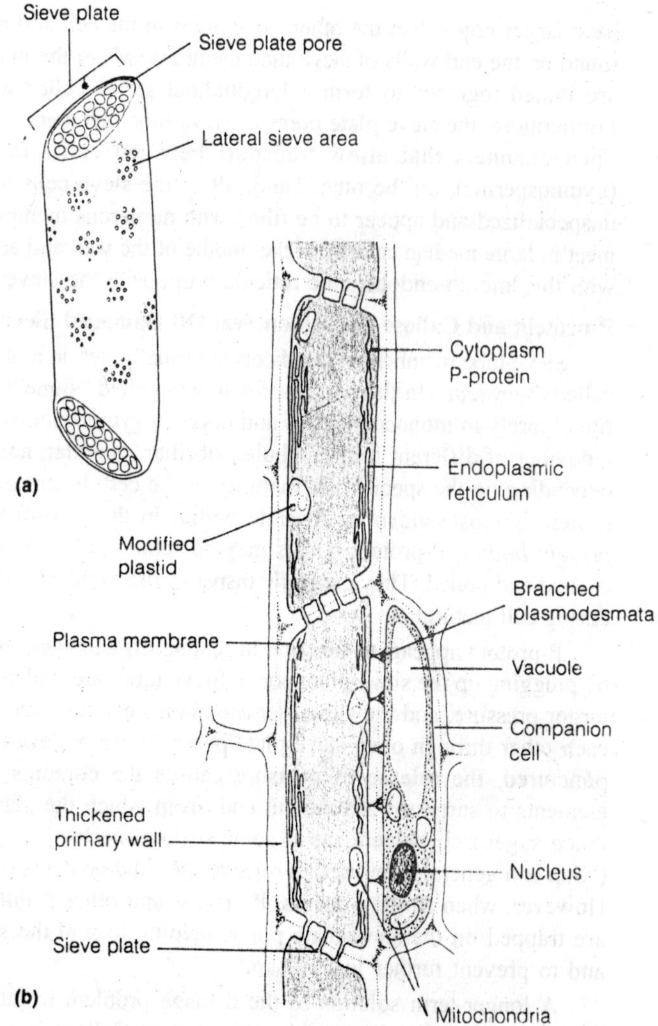

Fig. 5.3. Schematic drawings of mature sieve elements (sieve-tube members): (a) External view showing sieve plates and lateral sieve areas. (b) Longitudinal section showing two sieve-tube members joined together to form a sieve tube.

range in diameter from less than 1 μm to approximately 15 μm. The sieve areas of sieve-tube members (angiosperms) are more specialized than those of sieve cells (gymnosperms). For instance, some of the sieve areas of sieve-tube members are differentiated into *sieve plates*. Sieve plates have larger pores than the other sieve areas in the cell and are generally found on the end walls of sieve-tube members, where the individual cells are joined together to form a longitudinal series called a *sieve tube.* Furthermore, the sieve-plate pores of sieve-tube members are essentially open channels that allow transport between cells. In sieve cells (gymnosperms), on the other hand, all of the sieve cells are relatively unspecialized and appear to be filled with numerous membranes, which meet in large median cavities in the middle of the wall and are continuous with the smooth endoplasmic reticulum opposite the sieve areas.

P-protein and Callose Deposition Seal Off Damaged Sieve Elements

Sieve-tube members of the dicots are usually rich in a phloem protein called *P-protein.* (In the past, P-protein was called "slime"). P-protein is found rarely in monocot species and never in gymnosperms. It occurs in a number of different forms (tubular, fibrillar, granular, and crystalline) depending on the species and maturity of the cell. In immature cells, P-protein is most evident as discrete bodies in the cytosol known as *P-protein bodies*, P-protein bodies may be spheroidal, spindle-shaped, or twisted and coiled. They generally disperse into tubular or fibrillar forms during cell maturation.

P-protein appears to function in sealing off damaged sieve elements by plugging up the sieve-plate pores. Sieve tubes are under high internal turgor pressure, and the sieve elements in a sieve tube are connected to each other through open sieve-plate pores. When a sieve tube is cut or punctured, the release of pressure causes the contents of the sieve elements to surge toward the cut end, from which the plant count lose much sugar-rich phloem sap if some sealing mechanism did not exist. ("Sap is a general term used to refer to the fluid contents of plant cells.) However, when surging occurs, P-protein and other cellular inclusions are trapped on the sieve-plate pores, helping to seal the sieve element and to prevent further loss of sap.

A longer-term solution to the damage problem in most species is the production of *callose* in the sieve pores. Callose is a β-1 (1 → 3) glucan, which is identified by its reaction with aniline blue stain. Callose is synthesized by an enzyme in the plasma membrane and is deposited between the plasma membrane and the cell wall. The enzyme responsible for callose synthesis (callose synthetase) appears to be arranged vectorially in the plasma membrane, with substrate being supplied from the cytoplasmic side and product being deposited on the wall surface. It is now generally recognized that callose is synthesized in functioning sieve elements in response to damage and other stresses, such as mechanical stimulation and high temperatures. The deposition of such *wound callose* in the sieve pores efficiently seals off damaged sieve elements from surrounding intact tissue. As the sieve elements recover from damage, the callose disappears from these pores.

Callose is also found in sieve elements under circumstances other than wounding, although its function always seems to be one of sealing. *Definitive callose* is deposited in dying cells or sieve elements undergoing obliteration due to the formation of secondary tissues. Callose associated with dormancy is found in many overwintering plants, that is, perennial plants that have become dormant in the winter. Such *dormancy callose* is redissolved in the spring in preparation for resumption of transport and growth.

The Highly Specialized Sieve Elements are Functionally Supported by the Companion Cells

Each sieve-tube member is associated with one or more *companion cells*, the two cell types arising by the division of a single mother cell. Numerous intercellular connections, the *plasmodesmata*, penetrate the walls between sieve-tube members and their companion cells, suggesting a close functional relationship and ease of transport between the two cells. The plasmodesmata are often complex and branched on the companion-cell side. In the gymnosperms the companion-cell function is assumed by *albuminous cells*, which do not arise from the same mother cells as the sieve cells.

Companion cells are thought to take over some of the critical metabolic functions, such as protein synthesis, that are reduced or lost during the differentiation of the sieve elements The numerous mitochondria in companion cells may supply energy as ATP to the sieve element. In some species, photosynthetic products flow from the producing cells in the mesophyll through the companion cells to the sieve elements.

The transfer of materials to the sieve elements is further facilitated in some herbaceous dicots by the development of wall ingrowths in some companion cells or phloem parenchyma cells. Cells with such wall ingrowths are called *transfer cells*. The wall ingrowths greatly increase the surface area of the plasma membrane, thus increasing the potential for solute transfer across the membrane. Xylem parenchyma cell scan also be modified as transfer cells, probably serving to retrieve and reroute solutes moving in the xylem.

Patterns of Translocation

The direction of phloem transport is determined by the relative locations of the areas of supply and utilization of the products of photosynthesis. Phloem transport does not occur exclusively either in an upward or a downward direction and in not defined with respect to gravity. Translocation occurs from areas of supply (*sources*) to areas of metabolism or storage (*sinks*). Sources include any exporting organ, typically a mature leaf, that is capable of producing photosynthate in excess of its own needs. Another type of source is a storage organ during the exporting phase of its development. For example, the storage root of the biennial wild beet (*Beta maritima*) is a sink during the first growing season when it accumulates sugars received from the source leaves. During the second growing season the same root becomes a source; the sugars are remobilized and utilized to produce a new shoot, which ultimately becomes reproductive. In contrast, roots of the cultivated sugar beet (*Beta vulgaris*) can increase in dry mass during both the first and second growing seasons, so the leaves serve as sources during both flowering and fruiting stages. This occurs in cultivated varieties of beets because they have been selected for the capacity of their roots to act as sinks during all phases of development. Sinks include any non-photosynthetic organs of the plant and organs that do not produce enough photosynthetic products to support their own growth or storage needs. Roots, tubers, developing fruits, and immature leaves, which must import carbohydrate for normal development, are all examples of sink tissues. Both girdling and labeling studies support the source-to-sink pattern of translocation in the phloem.

Source-to Sink Pathways Follow Anatomical and Developmental Rules

While the overall pattern of phloem transport can be simply stated as source-to-sink movement, the specific pathways involved are often more complex, following anatomical and developmental rules. Sinks are not equally supplied by all source leaves on a plant; rather, certain sources preferentially supply specific sinks. In the case of herbaceous plant, the following generalizations can be made:

1. The proximity of the source to the sink is a significant factor. Thus, the upper mature leaves on a plant usually transport assimilates to the growing shoot tip and young, immature leaves while the lower leaves predominantly supply the root system. Intermediate leaves export in both directions, bypassing the intervening mature leaves.
2. The importance of various sinks may shift during plant development. Whereas the root and shoot apices are usually the major sinks during vegetative growth, fruits generally become the dominant sinks during reproductive development, particularly for adjacent and other nearby leaves.

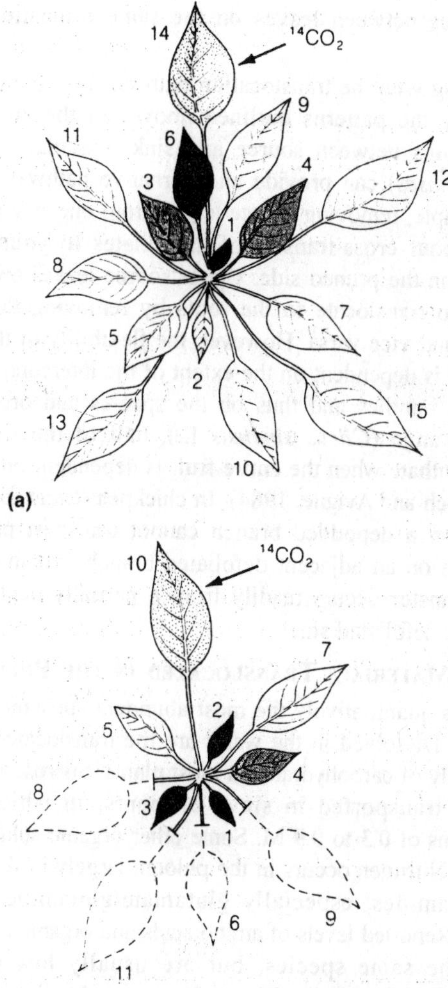

Fig. 5.4. Diagrams showing the distribution of radioactivity from a single labeled source leaf in an intact plant and in a pruned plant. (a) The distribution of radioactivity in leaves of an intact sugar beet plant (Beta vulgaris) one week after $^{14}CO_2$ was supplied for 4 hours to a single source leaf (arrow). (b) same as (a) except that all source leaves on the side of the plant opposite the labeled leaf were removed 24 hours prior to labeling.

3. Source leaves supply sinks with which they have direct vascular connections. A given leaf is generally connected via the vascular system to other leaves directly above or below it on the stem. Such a vertical row of leaves is called on *orthostichy*. The number of

internodes between leaves on the same orthostichy varies with species.

4. Interfering with the translocation pathway by wounding or pruning can alter the patterns outlined above. In the absence of direct connections between source and sink, vascular interconnections (*anastomoses*) can provide an alternative pathway. In sugar beet, for example, removing source leaves from one side of the plant can bring about cross-transfer of assimilates to young leaves (sink leaves) on the pruned side. Upper source leaves on a plant can be forced to translocate to the roots by removing the lower source leaves, and vice versa. However, the flexibility of the translocation pathway is dependent on the extent of the interconnections between vascular bundles and thus on the species and organs studied. In sweet orange (*Citrus sinensis* L.), half of the fruit receives no photosynthate when the entire fruit is dependent on a single source leaf (Koch and Avigne, 1984). In chickpea (*Cicer arietinum* L.), the leaves on a depodded branch cannot transport photosynthate to the pods on an adjacent defoliated branch, but in soybean plants, cross-transfer occurs readily from a partially depodded side to a partially defoliated side).

Materials Translocated in the Phloem

Water is quantitatively the most abundant substance transported in the phloem. Dissolved in the water are the translocated solutes, which consist mainly of carbohydrates in most plants. Sucrose is the sugar most commonly transported in sieve elements, in which it can reach concentrations of 0.3 to 0.9 M. Some other organic solutes also move in the phloem. Nitrogen occurs in the phloem largely in the form of amino acids and amides, especially glutamate/glutamine and aspartate/asparagine. Reported levels of amino acids and organic acids vary widely, even for the same species, but are usually low compared with carbohydrates. Almost all of the endogenous plant hormones, including auxin, gibberellins, cytokinins, and abscisic acid, have been found in sieve elements and are thought to be transported over long distances at least partly in these cells. Nucleotide-phosphates and enzymes have also been found in phloem sap. Inorganic solutes that move in the phloem include potassium, magnesium, phosphate, and chloride. In contrast, nitrate, calcium, sulfur, and iron are almost completely excluded form the phloem.

The discussion that follows begins with the methods used to identify materials translocated in the phloem. We will then examine the translocated sugars and the complexities of nitrogen transport in the plant.

Phloem Sap can be Collected and Analyzed

Since phloem tissue, like xylem tissue, consists of a mixture of conducting and nonconducting cells, identification of the material translocated in the sieve elements presents a formidable challenge. Methods based on labeling with radioactive isotopes and direct extraction of the tissue do not distinguish between compounds in transit and those outside the translocation pathway. However, the tendency of some species to exude phloem sap from wounds that sever sieve elements has been exploited to obtain relatively pure samples of the translocated material. The driving force for exudation is the high positive pressure in the sieve elements.

Most species do not exude detectable amounts of phloem sap because of the sealing mechanisms mentioned earlier. However, exudation rates can be increased by treating the cut surface with the chelating agent EDTA (ethylenediaminetetraacetic acid). Chelating agents inhibit callose formation by forming soluble complexes with divalent cations, particularly calcium, which is required by callose synthetase.

The major disadvantage of the wound-exudation method is that the fluid collected may not represent the true composition of the translocated material. Contaminants can originate from damaged parenchyma cells or even from the sieve elements themselves. Furthermore, the abrupt lowering of the turgor pressure in the sieve elements will, cause a decrease in the water potential of these cells. As a result, water from the surroundings will enter the sieve elements along a water potential gradient, causing a dilution of the phloem sap.

The ideal way to collect phloem sap would be to tap into a single sieve element by using a tiny syringe. Fortunately, nature has provided us with just such a probe, the aphid stylet! Aphids are small insects that feed by inserting their mouthparts, consisting of four tubular stylets, into a single sieve element of a leaf or stem. The high turgor pressure in the sieve elements force the cell contents through the insect's food canal and into the gut, where amino acids are selectively removed and some sugars are metabolized. The excess sap, still rich in carbohydrates, is then excreted as "honeydew," which can be collected for chemical analysis. Sap can also be collected from aphid stylets cut from the body of the insect after it has been anesthetized with CO_2. Exudate from severed stylets provides a more accurate picture of the substances present in the sieve elements than does honeydew, whose composition has been altered by the insect.

Aphid techniques have substantial advantages in the study of phloem physiology. Of particular importance, of course, is the fact that aphids tap a single sieve element, so there is no problem of contamination from other cell types. Exudation from severed stylets can continue for hours, which suggests that the aphid prevents normal sealing mechanisms from operating or that the pressure drop that occurs on penetration of the sieve elements is not sufficient to trigger sealing mechanisms.

The disadvantages of using aphids include the fact that sitting the insects at a desired location and severing the stylets without disrupting them require considerable patience and skill. In addition, aphids may induce reactions in the host plant by secreting saliva into the plant tissues. However, the magnitude of these reactions over the period of time necessary to collect sap for analysis is probably negligible.

The use of aphid stylets is thus the preferred technique for many laboratory experiments on herbaceous plants, for which a large, readily manipulated aphid species, *Longistigma caryae*, is available.

Sugars are Translocated in Nonreducing Form

Results from analyses of sap collected by the techniques outlined above indicate that the translocated carbohydrates are all nonreducing sugars. Reducing sugars, such as glucose and fructose, contain an exposed aldehyde or ketone group and can be assayed colorimetrically because of their ability to reduce Cu^{2+} to Cu^{+} in solution. In a nonreducing sugar the ketone or aldehyde group is reduced to an alcohol or combined with a similar group on another sugar. For example, the disaccharide sucrose is a nonreducing sugar composed of one glucose molecule (an aldose sugar) bound to one fructose molecule (a ketose sugar). Most researchers believe that the nonreducing sugars are the major translocates simply because they are less reactive than their reducing components.

Sucrose is the most commonly translocated sugar; many of the other mobile carbohydrates contain sucrose bound to varying numbers of galactose molecules. Raffinose consists of sucrose and one galactose molecule; stachyose, sucrose and two galactose molecules; and verbascose, sucrose and three galactose molecules. Translocated sugar alcohols include mannitol an sorbitol.

Transport Patterns of Nitrogenous Compounds in the Phloem and Xylem are Interrelated

Nitrogen is transported throughout the plant in either inorganic or organic form; the form that predominates depends on several factors, including the transport pathway. While nitrogen is transported in the

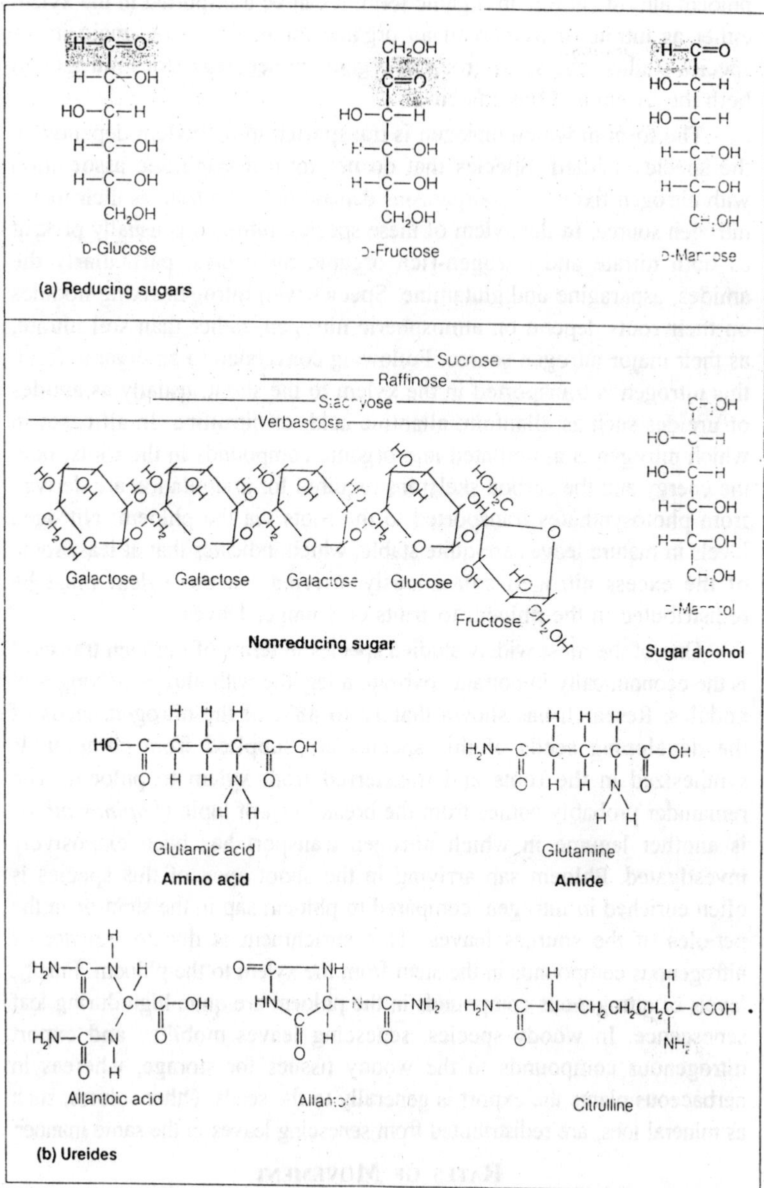

Fig. 5.5. Structures of compounds not normally translocated in the phloem and those commonly translocated in the phloem. (a) The reducing sugars glucose, fructose, and mannose are generally not found in phloem sap. (b) Compounds commonly translocated in the phloem.

phloem almost entirely in organic form, it can be transported in the xylem either as nitrate or as part of an organic molecule. In the majority of species studied, the same group of organic molecules carries nitrogen in both the xylem and the phloem.

The form in which nitrogen is transported in the xylem depends on the species studied. Species that do not form a symbiotic association with nitrogen-fixing microorganisms depend on soil nitrate as their major nitrogen source. In the xylem of these species, nitrogen is usually present as both nitrate and nitrogen-rich organic molecules, particularly the amides, asparagine and glutamine. Species with nitrogen-fixing nodules on their roots depend on atmospheric nitrogen, rather than soil nitrate, as their major nitrogen source. Following conversion to an organic form, this nitrogen is transported in the xylem to the shoot, usually as amides or ureides such as allantoin, allantoic acid, or citrulline. In all cases in which nitrogen is assimilated into organic compounds in the roots, both the energy and the carbon skeletons required for assimilation are derived from photosynthates transported to the roots via the phloem. Nitrogen levels in mature leaves are quite stable, which indicates that at least some of the excess nitrogen continuously arriving via the xylem must be redistributed in the phloem to fruits or younger leaves.

One of the most widely studied species in terms of nitrogen transport is the economically important soybean, a legume with nitrogen-fixing root nodules. Research has shown that 12 to 48% of the nitrogen needs of the developing seeds of this species are supplied from compounds synthesized in the roots and transferred from xylem to phloem. The remainder probably comes from the breakdown of lupin (*Lupinus albus*) is another legume in which nitrogen transport has been extensively investigated. Phloem sap arriving in the shoot apex of this species is often enriched in nitrogen, compared to phloem sap in the stem or in the petioles of the sources leaves. This enrichment is due to transfer of nitrogenous compounds in the stem from the xylem to the phloem. Finally, levels of nitrogenous compounds in the phloem are quite high during leaf senescence. In woody species, senescing leaves mobilize and export nitrogenous compounds to the woody tissues for storage, whereas in herbaceous plants the export is generally to the seeds. Other solutes, such as mineral ions, are redistributed from senescing leaves in the same manner.

Rates of Movement

The rate of movement of material is in the sieve elements can be expressed in two ways: as *velocity*, the linear distance traveled per unit time, or as *mass transfer rate*, the quantity of material passing through a given cross section of phloem or sieve elements per unit time. Mass transfer rates based on the cross-sectional area of the sieve elements are preferred, since the sieve elements are the actual conducting cells of the phloem.

In the original publications the units of velocity were in centimeters per hour, while the units of mass transfer were in grams per hour per square centimeter of phloem or sieve elements. The currently preferred units (SI units) are meters (m) or millimeters (mm) for length, second (s) for time, and kilogram (kg) for mass.

Velocities of Phloem Transport Average One Meter per Hour

Both velocities and mass transfer rates can be measured with radioactive tracers, and this is the technique most commonly used to determine transport velocities. In the simplest type of experiment, ^{11}C- or ^{14}C-labeled CO_2 is applied for a brief period of time to a source leaf (pulse labeling), and the arrival of label at a sink tissue or at some point along the pathway is monitored with an appropriate detector. The length of the translocation pathway divided by the time interval required for label to be first detected at the sink yields a measure of velocity, at least for the fastest-moving labeled component. A more accurate measurement of velocity is obtained by monitoring the arrival of label at two points along the pathway, since this method excludes from the time measurement the time required for fixation of labeled carbon by photosynthesis, for conversion into transport sugar, and for accumulation of sugar in the sieve elements of the source leaf. In general, velocities measured by a variety of techniques average about 1 m h^{-1} and range from 30 to 150 cm h^{-1}, with a few studies reporting speeds an order of magnitude greater than these. The variability in the measured velocities is due to species differences, differences between experimental methods, and problems inherent in the experimental methods. Despite this variability, it is clear that transport velocities in the phloem are quite high and well in excess of the rate of diffusion over long distances. Any proposed mechanism of phloem translocation must account for these high velocities.

Phloem Loading

Several transport steps are involved in the movement of photosynthetic products (photosynthate) from the mesophyll chloroplasts to the sieve elements of mature leaves:

1. In the typical case of a plant that translocates sucrose, triose phosphate formed during photosynthesis must first be transported from the chloroplast to the cytosol, where it is converted to sucrose.
2. Sucrose then moves from the mesophyll cell to the vicinity of the sieve elements in the smallest veins of the leaf. This pathway usually

involves a distance of only two or three cell diameters and is referred to as the *short-distance transport* pathway.

3. In the third step, *phloem loading*, sucrose is actively transported into the sieve elements.

Once inside the sieve elements, sucrose and other solutes are translocated away from the source, a process known as *export*. Translocation through the vascular system to the sink is referred to as *long-distance transport*.

The process of phloem loading at the source, and unloading at the sink, are believed to produce the driving force for translocation and thus are of considerable basic as well as agricultural importance. Once the mechanisms are understood, it may be possible to increase crop productivity by increasing the accumulation of photosynthate by edible sink tissues, such as cereal grains.

Phloem Loading of Sugar Requires Metabolic Energy

In source leaves, sugars become more concentrated in the sieve elements and companion cells than in the mesophyll. This difference in solute concentration can be demonstrated by measuring the osmotic potential of the various cell types in the leaf, using a combination of plasmolysis and light or electron microscopy. In sugar beet, for example, the osmotic pressure of the mesophyll is approximately 1.3 MPa, while the osmotic pressure of the sieve elements and companion cells are on the order of 3.0 Mpa. Most of this difference in osmotic pressure is thought to the due to sugar, and specifically to sucrose, since sucrose is the major transport sugar in this species. Furthermore, ^{14}C-labeled sucrose supplied exogenously to a source leaf of sugar beer accumulates in the sieve elements and companion cells of the major veins, as does sucrose derived from $^{14}CO_2$. The fact that sucrose, an uncharged solute, is at higher concentration in the sieve element-companion cell complex than in surrounding cells indicates that sucrose is transported against its chemical potential gradient and is evidence for active transport of this solute. Other data support the concept that phloem loading requires metabolic energy. For example, treating source tissues with respiratory inhibitors decreases the ATP concentration in the tissue and also inhibits the loading of exogenous sugar.

The Pathway from the Mesophyll Cell to the Sieve Elements is at Least Partly Apoplastic

We have seen that solutes (mainly sugars) must move in source leaves from the photosynthesizing cells in the mesophyll to the veins. Sugars might move entirely through the symplast via the plasmodesmata

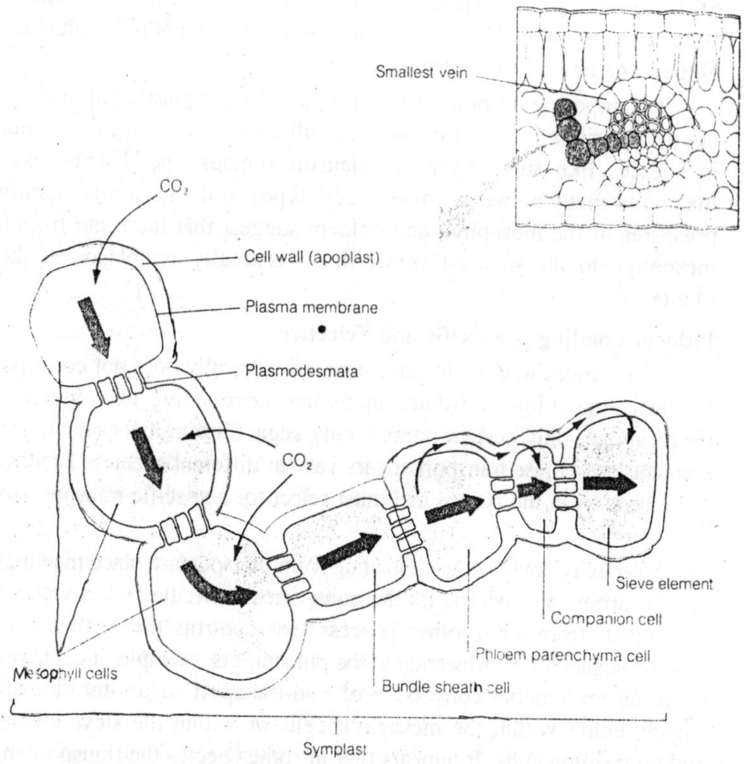

Fig. 5.6. Diagram of possible pathways of phloem loading in source leaves.

or, alternatively, might enter the apoplast at some point en-route to the phloem. In the latter case, the sugars could be actively loaded from the apoplast into the sieve elements and companion cells by an energy-driven carrier located in the plasma membranes of these cells.

The exit of sucrose from mesophyll cells into the apoplast is at least partly controlled by the level of certain substances, such as potassium, in the apoplast. A high level of potassium in the apoplast of sugar beet source leaves increases the rate at which sucrose enters the apoplast, thus coordinating a favourable nutrient supply, increased translocation to sinks, and enhanced sink growth. The mechanism of sucrose efflux into the apoplast is unknown. However, loading of sucrose from the apoplast to the sieve elements is thought to be regulated by osmotic pressure or, more likely, the turgor pressure of the sieve elements. According to this model, a decrease in sieve-element turgor below a certain point would lead to a compensatory increase in loading. The entry of sucrose into the sieve elements is also affected by sucrose concentration, with higher concentrations in the apoplast increasing phloem loading by the sucrose carrier.

It is important to point out that a partially apoplastic pathway does not necessarily hold true for loading in all species or even for all solutes in species like sugar beet. In cucurbit species, the distribution of plasmodesmata between various cell types and the similar osmotic pressures of the mesophyll and phloem suggest that the route from the mesophyll to the smallest vein phloem is totally symplastic in these plants.

Phloem Loading is Specific and Selective

The composition of sieve element sap generally does not correspond to the composition of solutes in tissues surrounding the phloem. As noted above, only certain nonreducing sugars are translocated in sieve elements, and these transport sugars vary in different species. Therefore, some process in the source leaf must select for a specific transport sugar to sugars.

Selectivity could occur in loading from the apoplast, since membrane-bound carriers are specific for the transported molecule. Selectivity could also result from some other process that controls the availability of specific sugars for translocation in the phloem. For example, it could result from the preferential conversion of non-transport sugars into transport sugars, either within the mesophyll cells or within the sieve elements and companion cells. It appears that in some species the transport sugar is selected by the carrier for loading from the apoplast, whereas in other species it is selected by processes other than the loading system.

Not all Substances Transported in the Phloem are Actively Loaded into the Sieve Elements

Many substances, such as organic acids and plant hormones, are found in the phloem sap at lower concentrations than the carbohydrates. These substances are probably not actively loaded into the sieve element–companion cell complex. They may be taken up directly by diffusion across the phospholipid bilayer of the plasma membrane of the sieve element–companion cell complex or by a passive carrier in the plasma membrane of those cells, or they may diffuse into the sieve elements via the symplast. Once in the sieve elements, they are swept along in the translocation stream by bulk flow, the motive force being generated by the active loading of only certain sugars or amino acids. Many substances not normally found in plants, such as herbicides and fungicide, can be transported in the phloem because of their ability to diffuse through membranes at an intermediate rate. In other words, they diffuse through membranes rapidly enough to allow considerable accumulation in the sieve elements, but slowly enough that they do not diffuse out completely before reaching a sink tissue. Substances that are not transported in the phloem (such as calcium) apparently cannot enter the sieve elements.

Sucrose Loading is Driven by a Proton Gradient Generated at the Expense of ATP

Most active transport mechanisms use metabolic energy supplied in the form of ATP. The hydrolysis of ATP can be coupled to the movement of solutes across membranes in at least two ways. In *primary active transport* the same membrane protein is thought to break down ATP and use the energy released to move the solute across the membrane barrier against a chemical potential gradient. In plant cells, the primary active transport ATPase is a proton pump. In secondary transport, solute movement against a chemical potential gradient is driven not directly by ATP hydrolysis, but indirectly by the proton gradient established by the primary active transport ATPase. The high proton concentration in the apoplast and the spontaneous tendency toward equilibrium (equal apoplast and symplast) cause protons to diffuse back into the symplast. Specific carrier molecules couple this movement to the transport of sucrose. This type of transport is known as sucrose/proton symport or co-transport.

A greater deal of research on phloem loading has been done on relatively few species, such as sugar beet and corn. In these plants and a few others, the present data support the pathway for loading. In corn leaves, sugars probably diffuse in the symplast from the mesophyll to the vicinity of the minor-vein sieve elements. In this species the movement of water and dissolved solutes through the apoplast is blocked by suberized cell walls in the bundle sheath. The sugars then enter the apoplast by a largely unknown mechanism. In sugar beet leaves, sugars might initially follow a similar symplastic pathway, or they might enter the apoplast over the entire mesophyll and diffuse to the minor veins in the apoplast; the actual initial pathway is not known. In both species the sugars are then actively and selectively loaded from the apoplast into the sieve elements and the companion cells by a carrier present in their plasma membranes, giving rise to the high sugar concentrations present in these cells. The carrier mechanism appear to be a sucrose/ proton symport. As we shall see, such active phloem loading plays a critical role in the mechanism of phloem translocation most widely accepted today.

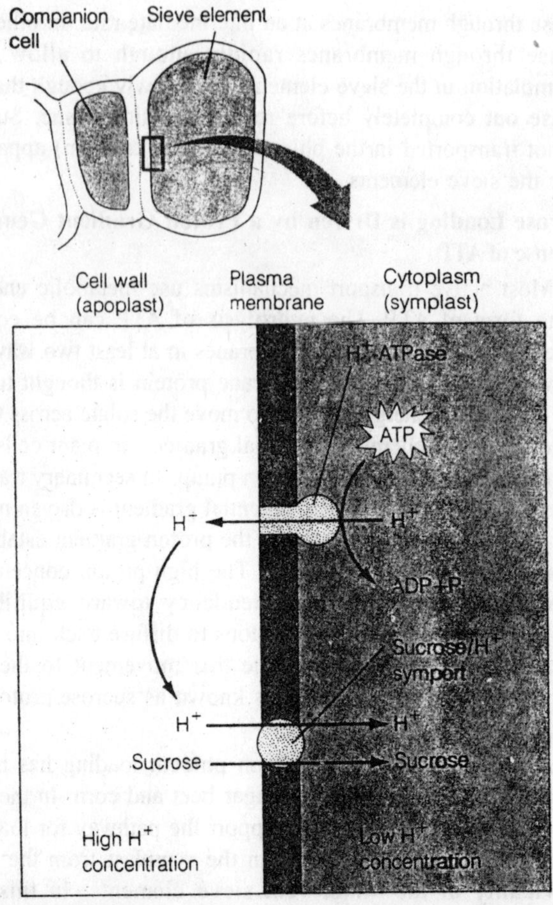

Fig. 5.7. In the cotransport model of sucrose loading into the symplast of the sieve element–companion cell complex, the plasma membrane ATPase pumps protons out of the cell into the apoplast, establishing a high proton concentration there. The energy in this proton gradient is then used to drive sucrose transport into the symplast of the sieve element–companion cell complex through a sucrose/H^+ symporter.

Phloem Unloading and Sink-to-Source Transition

The detailed discussion above illustrates the events leading up to export of sugars from sources. Phloem *unloading* is the process by which translocated sugars exit from the sieve elements of sink tissues.

Translocation into sink organs, such as developing roots, tubers, and reproductive structures, is termed import. In a sense, the events following import are simply the reverse of the steps in sources. The initial step is the unloading of sugars from the sieve elements. After unloading, the sugars are transported to cells in the sink for storage or metabolism.

Phloem Unloading and Transport into Receiver Cells can be Symplastic or Apoplastic

In vegetative sinks that are growing, such as roots and young leaves, unloading and transport into receiver cells are usually symplastic. Both structural observations and experimental results support this hypothesis. In other sink organs unloading is apoplastic. In storage sinks, such as sugar beet root and the sugarcane (*Saccharum officinarum* L.) stem, sucrose is unloaded into the apoplast prior to entering the symplast of the sink. In reproductive sinks (developing seeds), an apoplastic step is necessary because there are no symplastic connections between the maternal tissues and the tissues of the embryo.

When unloading is symplastic, transport sugar moves through the plasmodesmata to the receiver cell, where it can be metabolized in the cytosol or the vacuole before entering metabolic pathways associated with growth of the tissue. When unloading is apoplastic, the transport sugar can be partially metabolized in the apoplast itself. For example, in sugarcane stem and corn kernels, sucrose is split into its components in the apoplast by an enzyme called invertase and is taken up as glucose or fructose. In sugar beet root and soybean seeds, on the other hand, sucrose crosses the apoplast unchanged. Storage sinks can accumulate sucrose in the vacuole, or the absorbed sugars can be metabolized to some other solute before storage.

The information emphasizes the variety of transport mechanisms involved in unloading. The patterns discussed here are only general ones and should not be taken as hard-and-fast rules. Rather, for any particular species or organ, the operative pathway must be determined experimentally.

Transport into Sink Tissues Depends on Metabolic Activity

Numerous studies with inhibitors have shown that transport into sink tissues is energy dependent. The site or sites at which energy is required vary with the species or organ studied. In the case of apoplastic unloading, sugars must cross at least two membranes: the membrane of the sieve element-companion cell complex and the membrane of the receiver cell. When transport into the vacuole of the sink cell occurs, the tonoplat must also be traversed. Carriers are thought to function in transport across these membranes, and at least one of the membrane transport steps has been shown to be active (dependent on metabolism) in sinks where unloading is apoplastic. In sinks where unloading is symplastic, no membranes are crossed during uptake into the sink cells.

Unloading through the plasmodesmata is passive, since transport sugars move from a high concentration in the sieve elements to a low concentration in the sink cells. The low concentration in the sink cells is maintained by respiration and by conversion of the transport sugars to other compounds needed in growth. Metabolic energy is required directly in these growing sinks for respiration and biosynthetic reactions and only indirectly for nutrient uptake. Metabolic conversions also help to maintain the concentration gradient for sink uptake in some organ with apoplastic unloading.

The Active Transport step during Apoplastic Unloading Depends on the Species and Organ

Storage organs with apoplastic unloading, such as sugar beet root and sugarcane stem, often accumulate sugars to high concentrations. This fact is consistent with active membrane transport, since energy is generally required to move sugars into storage compartments against a concentration gradient. However, it is important to note that the actual unloading step is probably passive in both these sinks; that is, movement of sucrose out of the sieve elements into the apoplast occurs down a concentration gradient, from a high to a low concentration. A low apoplastic sucrose level is maintained in sugarcane by the cell wall invertase, which splits sucrose into glucose and fructose; in the case of sugar beet storage root, a subsequent active transport of sucrose at the tonoplast of receive cells maintains the concentration gradient.

Developing seeds have proved to be a most interesting system in which to study unloading processes. In legumes such as soybean, the embryo can be removed from the seedcoat. Thus, unloading from the seedcoat into the apoplast can be studied without the influence of the embryo, and uptake into the embryo can also be investigated separately. Such studies have shown that in these species, at least part of the uptake into the embryo is active, carrier-mediated transport and that the entry of sucrose into the apoplast is probably carrier-mediated and may require metabolic energy.

In general, sugar/proton symport mechanisms appear to function in retrieval from the apoplast (for instance, in sucrose uptake in the soybean embryo and hexose uptake in sugarcane storage cells). Sugar transport into the vacuoles of storage cells such as those of sugarcane and sugar beet is accomplished by a *sucrose/proton antiport*. In this case, an ATPase pumps protons into the vacuole; the antiport carrier moves sucrose into the vacuole in exchange for protons, which exit and vacuole down their electrochemical potential gradient.

The Transition from Sink to Source is a Gradual Developmental Process in Leaves

Dicot leaves begin their development as sink organs. A transition from sink to source status begins later in development, generally when the leaf is approximately 25% expanded, and is usually complete when the leaf is 40 to 50% expanded. Export from the leaf begins at the tip or apex of the blade and progresses toward the base until the whole leaf is exporting. During the transition period, the tip exports sugar while the base imports it from the other source leaves, as shown by autoradiography. What causes import to cease and export to begin? The maturation of leaves is accompanied by a large number of functional and anatomical changes, many of which are simply preparatory for the beginning of export. Export actually begins when phloem loading has accumulated sufficient photosynthate in the sieve elements to drive translocation out of the leaf.

The cessation of import and the initiation of export are separate events in the development of a leaf. In albino leaves of tobacco, which have not chlorophyll and therefore are incapable of photosynthesis, import stops at the same developmental stage as in green leaves, even though export is not possible. Therefore some other change must occur in developing leaves that causes them to cease importing sugars from other leaves. Such a change could involve blockage of the unloading pathway at some point in the development of mature leaves. Since unloading in sink leaves appears to be symplastic in nature, plasmodesmatal closure could account for the cessation of unloading and import into transition leaves. With the unloading pathway blocked, the sieve elements could accumulate enough sugar for export to be initiated in normal leaves. Additional data support the hypothesis that the unloading pathway is blocked in mature leaves. Some translocation and import of labeled sugars into a mature leaf (from other mature leaves!) can be induced by rather extreme conditions, such as darkening of the treated leaf and elimination of any alternative sinks. However, the imported sugars are not unloaded: they remain in the veins, indicating that the symplastic pathway to the mesophyll has been blocked during the maturation of the leaf.

Mechanism of Phloem Translocation

The mechanism of phloem translocation was a subject of research from the 1930s to the mid-1970s. Now, one theory is generally accepted as the correct explanation for translocation. This theory, called the *pressure-flow hypothesis*, accounts for most of the experimental and structural data currently available for angiosperms.

The discussion here will begin with a description of the various types of models proposed in the past to account for phloem translocation. A second section will deal with the considerable evidence in favour of the pressure-flow hypothesis in angiosperms. The discussion will end with a note of caution: pressure flow may not be a complete description of translocation in all species.

Active and Passive Mechanisms have been Proposed to Account for Phloem Translocation

Theories of phloem translocation can be categorized as active or passive. Both types of theories assumes that energy is required for loading in the source and for sink uptake. The active theories further assume that an additional expenditure of energy is required to drive the translocation process in the sieve elements, whereas in the passive theories, energy is required only to maintain the functional integrity of the sieve elements, not to drive translocation itself.

One of the original active theories was that solutes were carried from one end of a sieve elements to the other by cytoplasmic streaming or cyclosis, with transfer occurring across the sieve plate by some unknown mechanism. This theory had to be abandoned because cytoplasmic streaming has never been observed in mature, functioning sieve elements. In addition, there is no evidence that sieve elements contain actin microfilaments, which function in cytoplasmic streaming in other plant cells. Another early active model for phloem translocation suggested that P-protein might provide the motive force for solute movement by some type of contractile or peristaltic action, analogous to the action of actin microfilaments in muscle or of microtubules in cilia and flagella. However, there is no strong evidence that P-protein is similar to either actin or tubulin.

A second type of active theory involved energy-driven transport of solutes from one sieve element to another across the sieve plate. Such active transport of solutes was thought to be required because electron micrographs of phloem tissue appeared to show that functional sieve-plate pores were normally blocked by callose and P-protein. If the sieve-plate pores between sieve elements were blocked when functional, passive models of translocation based on bulk flow of solutes would be impossible. As we shall see, the apparent plugging of sieve-plate pores observed in these early electron micrographs was an artifact of the fixation method. The sieve-plate pores are actually open when functional, so it is unnecessary to invoke an active mechanism of solute transport across sieve plates.

According to the Pressure-Flow Hypothesis, Translocation in the Phloem is driven by a Pressure Gradient from Source to Sink

The passive theories of translocation include diffusion and the pressure-flow hypothesis. Diffusion is far too slow to account for the velocities of solute movement observed in the phloem. Translocation velocities average 1 meter per hour, while the rate of diffusion is 1 meter per eight years.

The pressure-flow hypothesis on the other hand, is widely accepted as the most probable mechanism of phloem translocation. First proposed by Ernst Munch in 1930, the pressure-flow hypothesis states that the flow of solution in the sieve elements is driven by an osmotically generated pressure gradient between source an sink. The pressure gradient is established as a consequence of phloem loading at the source and phloem unloading at the sink. That is, energy-driven phloem loading generates a high osmotic pressure in the sieve elements of the source tissue, causing a steep drop in the water potential. In response to the water potential gradient water enters the sieve elements and causes the turgor pressure to increase. At the receiving end of the translocation pathway, phloem unloading leads to a lower osmotic pressure in the sieve elements of sink tissues. As the water potential of the phloem rises above that of the xylem, water tends to leave the phloem in response to the water potential gradient, causing a decrease in the turgor pressure of the phloem sieve elements of the sink.

If no cross walls were present in the translocation pathway—that is, if the entire pathway were a single membrane-bound compartment—the two different pressures at the source and sink would rapidly come to near equilibrium. The presence of sieve plates greatly increases the resistance along the pathway and results in the generation and maintenance of a substantial pressure gradient in the sieve elements between source and sink. The sieve element contents are physically pushed along the translocation pathway by bulk flow, much like water flowing through a garden hose. Note that this model implies that some of the water circulates throughout the plant between the transpiration (xylem) and translocation (phloem) pathways.

From close inspection of the water potential values shown in, it is apparent that water in the phloem is moving up a water potential gradient from source to sink. Such water movement does not transgress the laws of thermodynamics, however, since the water is moving by bulk flow rather than by osmosis. That is, no membranes are crossed during transport from one sieve tube to another and solutes are moving at the

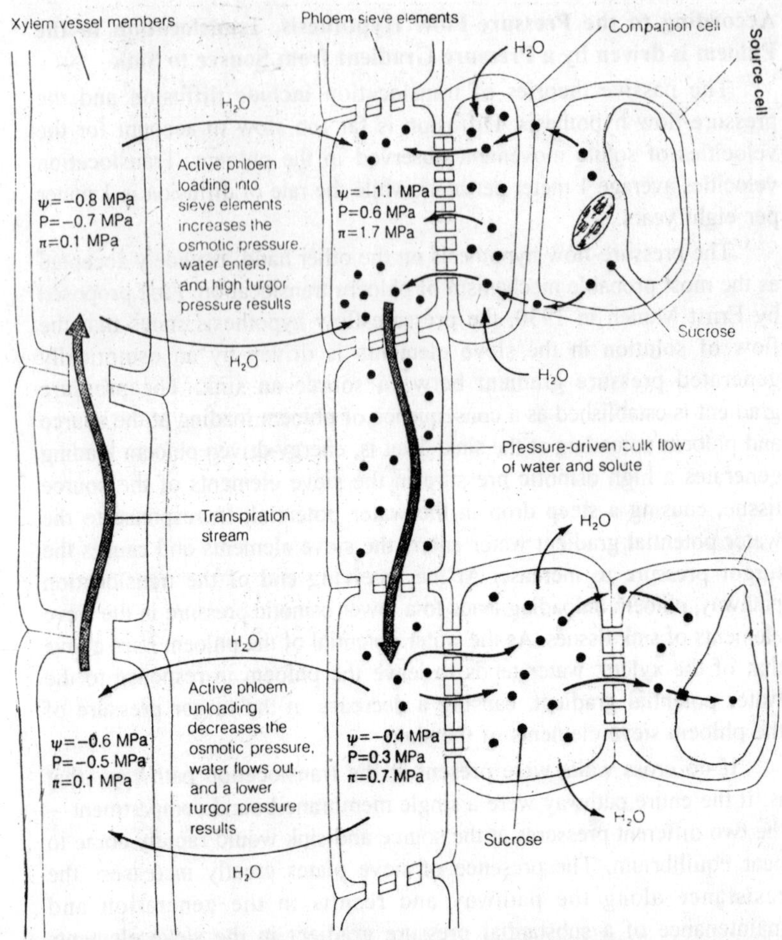

Fig. 5.8. Schematic diagram showing the pressure-flow model. Possible values for ψ, P, and π in the xylem and phloem are illustrated.

same rate as the water molecules. Under these conditions, the osmotic pressure, π, cannot contribute to the driving force for water movement, although it still influences the water potential. Water movement in the translocation pathway is therefore driven by the pressure gradient rather than the water potential gradient. Of course, the passive, pressure-driven long-distance translocation in the sieve tubes ultimately depends on the active short-distance transport mechanisms involved in phloem loading and unloading. These active mechanisms are responsible for setting up the pressure gradient in the first place.

Predictions of the Pressure-Flow Model

Since the pressure-flow hypothesis has gained such wide acceptance, the predictions based on the model demand special attention. First of all, the sieve-plate pores must be unobstructed. If P-protein or other materials blocked the pores, the resistance to flow of the sieve-element sap would be too great. Second, true *bidirectional transport* (i.e. transport in both directions) in a single sieve elements cannot occur. A mass of solution precludes such bidirectional movement, since a solution can flow in only one direction in a pipe! Bidirectional movement of solutes within the phloem can occur, but in different vascular bundles or in different sieve elements. Third, since the pressure-flow hypothesis is a passive theory, great expenditures of energy would not be required in order to drive translocation in the tissues along the path. Energy would be required in the path only to maintain the structure of the sieve elements and the integrity of the cell membranes. Therefore, treatments that restrict the supply of ATP, such as low temperature, anoxia, and metabolic inhibitors, should not stop translocation when applied to the path tissues. Note, however, that energy dependence of the translocation mechanism is not demonstrated if the inhibitor halts translocation by disrupting cell and membrane structure and not solely by reducing ATP supply. The cell membrane must remain intact and functional in order to retain solutes in the transport stream, and the sieve-plate pores must remain open for translocation to proceed. Fourth, the pressure-flow hypothesis demands the presence of a positive pressure gradient. Turgor pressure must be higher in sieve elements of sources than in sieve element of sinks, and the pressure difference must be large enough to overcome the resistance of the pathway and to maintain flow at the observed velocities.

Many attempts have been made to test the predictions of pressure-flow model. Often the results, or at least the interpretations given them, have been contradictory. Emphasis will be given here to experiments whose results can be interpreted in a relatively straightforward fashion. The following is a summary of the evidence supporting this hypothesis.

The Sieve-Plate Pores are Essentially Open Channels Connecting One Sieve Tube Member to Another

The ultrastructure of sieve elements is not easy to investigate because of their high internal pressure. When the phloem is excised or fixed (killed) slowly with chemical fixatives, the turgor pressure in the sieve elements is released. The contents of the cell, including P-protein, surge toward the point of pressure release, and in the case of sieve-tube members, accumulate on the sieve plates. That is probably why many of the earlier electron micrographs show sieve plates that are obstructed. As mentioned above, if the pores of the sieve plates were normally blocked, translocation by pressure flow could not occur. Now, rapid freezing and fixation techniques are available that permit a fairly reliable picture of undisturbed sieve elements to be obtained. Electron micrographs of sieve-tube members prepared by such techniques indicate that P-protein is normally found along the periphery of the sieve-tube members, or evenly distributed throughout the lumen of the cell. Furthermore, the pores contain P-protein in similar positions, lining the pore or in a loose network. The open condition of the pores has been observed in a number of angiosperm species, including cucurbits, sugar beet, and bean, consistent with the pressure-flow model.

Bidirectional Transport in a Single Sieve Element has not Been Demonstrated

Bidirectional transport has been investigated by applying two different tracers to two source leaves, one above the other. Each source receives one of the tracers, and some point in between the two sources is monitored for he presence of both tracers. Alternatively, a single tracer is applied to an internode and detected at points above and below the site of application. Detection techniques depend on the tracer used and include autoradiography and visualization of fluorescent dyes in fresh sections. In some experiments, aphids are used to collect exudate from a single sieve element, and the exudate is analyzed.

Transport in two direction has been often detected in sieve elements of different vascular bundles in stems. Transport in two directions has also been seen in adjacent sieve elements of the same bundle in petioles. This can occur in the petiole of a leaf that is undergoing the transition from sink to source and simultaneously importing and exporting assimilates through its petiole. However, bidirectional transport in a single sieve element has never been convincingly demonstrated. Experiments that purport to demonstrate such bidirectional transport can be interpreted in other ways.

The Rate of Translocation is Relatively Insensitive to the Energy Supply of the Path Tissues

In plants, such as sugar beet, that are able to survive periods of low temperatures, rapidly chilling a short segment of a source-leaf petiole to approximately 1°C does not cause a sustained inhibition of mass transport out of the leaf. Rather, there is a brief period of inhibition, after which transport returns to the control rate in one to several hours after the beginning of the cold treatment. These changes are not most easily seen

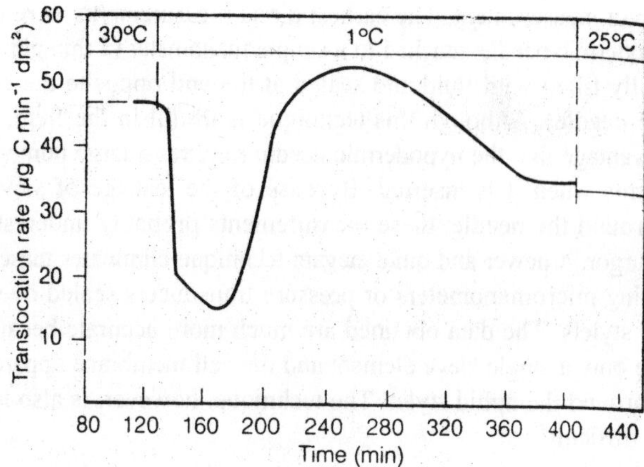

Fig. 5.9. The response of translocation in sugar beet (Beta vulgaris) to loss of metabolic energy by chilling the tissue.

in time courses obtained using steady-state labeling. Chilling reduces the respiration rate (and ATP production) by about 90% in the petioles at a time when translocation has recovered and is proceeding normally. In fact, calculations for sugar beet indicate that for each molecule of ATP generated in a chilled sieve element-companion cell complex, approximately 600,000 molecules of sucrose are transported through the element. Similar results are obtained when squash (*Cucurbita melopepo torticollis*) petioles are treated with an atmosphere consisting of 100% nitrogen in the dark; transport persists even though anaerobic metabolism is the only source of ATP throughout a significant portion of the transport pathway. The main point here is that the energy requirement for transport through the pathway of these plants is small, consistent with the pressure-flow hypothesis.

In the case of chilling-sensitive plants, such as bean (*Phaseolus vulgaris* L.), chilling the petiole of a source leaf to 10°C does inhibit translocation out of the leaf. Treating the petiole with a metabolic inhibitor (cyanide) also inhibits translocation in this plant. However, examination of the treated tissue by electron microscopy reveals blockage of the sieve-plate pores by cellular debris in both cases. Clearly, these results do not bear on the question of whether energy is required for translocation along the pathway.

Pressure Gradients in the Sieve Elements are sufficient to Drive a Mass Flow of Solution

Turgor pressure in sieve elements can be determined either by calculation from the water potential and osmotic pressure ($P = \psi + \pi$) or by direct measurement. The earliest direct measurements were made with a hypodermic needle attached to a simple manometer (a thin glass capillary partially filled with fluid and sealed at the end opposite the connection to the needle). Although this technique is useful in the field, it has the disadvantage that the hypodermic needle ruptures a large number of sieve elements when it is inserted. Because of the leakage of sieve-element sap around the needle, these measurements probably underestimate the true turgor. A newer and quite elegant technique eliminates these problems by using micromanometers or pressure transducers sealed over exuding aphid stylets. The data obtained are much more accurate because aphids pierce only a single sieve element and the cell membrane apparently seals well around the aphid stylet. The technique, however, is also technically quite difficult!

When sieve-element turgor is measured by the techniques described above, the pressure at the source is generally found to be higher than that at the sink. For example, a pressure difference of 0.11 MPa was detected between the source and sink tissues of squirting cucumber (*Ecballium elaterium*) by the hypodermic needle/manometer method. In soybean, the observed pressure difference between source and sink has been shown to be sufficient to drive a mass flow of solution through the pathway, taking into account the path resistance (mainly caused by the sieve-plate pores), the path length, and the velocity of translocation. The actual pressure difference between source and sink was calculated from the water potential and osmotic pressure to be 0.41 MPa, and the pressure difference required for translocation by pressure flow was calculated to be 0.12 to 0.46 MPa. Thus, the observed pressure difference appears to be sufficient to drive mass flow through the phloem.

All the experiments and data described above support the operation of pressure flow in angiosperm phloem. The lack of an energy requirement in the pathway and the presence of open sieve-plate pores are definitive evidence for a passive mechanism. The failure to detect bidirectional transport or motility proteins, as well as the positive data on pressure gradients, are all in accord with the pressure-flow hypothesis.

The Mechanism of Phloem Transport in Gymnosperms may be Different from that in Angiosperms

While pressure flow is adequate to explain translocation in angiosperms, it may not be sufficient in gymnosperms. Very little physiological information on gymnosperm phloem is available, and speculation about translocation in these species is based almost entirely on ultrastructural data. The sieve cells of gymnosperms are similar in many respects to sieve-tube members of angiosperms. However, the sieve areas of sieve cells are relatively unspecialized and do not appear to consist of open pores. Rather, the pores are filled with numerous membranes that are continuous with the smooth ER adjacent to the sieve areas. Such pores are clearly inconsistent with the requirements of the pressure-flow hypothesis. Either the picture presented by these electron micrographs is an artifact of the fixation process, or translocation in gymnosperms involves some other mechanism, perhaps with the endoplasmic reticulum playing a significant role.

Assimilate Allocation and Partitioning

The photosynthetic rate determines the total amount of fixed carbon available to the leaf. However, the amount of fixed carbon available for translocation depends on subsequent metabolic events. The regulation of the diversion of fixed carbon into the various metabolic pathways is termed *allocation.*

As we have seen, once the transport sugar enters the sieve element, it and the solvent in which it is dissolved move by mass flow along a pressure gradient in the direction of a sink. The vascular bundles form a system of pipes that can direct the flow of photoassimilates to the various organs: young leaves, stems, roots, fruits, or seeds. However, the vascular system is often highly interconnected, forming an open network that allows source leaves to communicate with multiple sinks. Under these conditions, what determines the volume of flow to any given sink? The differential distribution of photoassimilates within the plant is termed *partitioning*.

Allocation Includes the Storage, Utilization, and Transport of Fixed Carbon in the Plant

1. *Synthesis of storage compounds.* Starch is synthesized and stored within chloroplasts and, in most species, is the primary storage form that is mobilized for translocation during the night period. Such plants are called starch storers. In some organs of certain grasses, fructans (polymerized fructose molecules) are the storage compounds, rather than starch.
2. *Metabolic utilization.* Fixed carbon can be utilized within various compartments of the photosynthesizing cell to meet the energy needs of the cell or to provide carbon skeleton for the synthesis of other compounds required by the cell.
3. *Synthesis of transport compounds.* Fixed carbon can be incorporated into transport sugars for export to various sink tissues. A portion of

the transport sugar can also be stored temporarily in the vacuole. In most species studied, this stored sucrose is highly transitory, providing a buffer against short-term changes in sucrose synthesis. However, in some species, such as barley (*Hordeum vulgare* L.), fixed carbon is stored for use during the night period primarily as sucrose, with very little being stored as starch. Such plants are called sucrose storers.

Allocation is a key process in sink tissues as well. Once the transport sugars have been unloaded and enter the receiver cells, they can remain as such or can be transformed into various other compounds. In storage sinks, fixed carbon can be accumulated as sucrose or hexose in vacuoles or as starch in amyloplasts. In growing sinks, sugars can be utilized for respiration and for the synthesis of other molecules required for growth.

Once Synthesized, Transport Sugars are Partitioned among the Various Sink Tissues

The cellular processes involved in allocation largely determine the relative strength of the various sinks. That is, the greater the ability of a sink to store or metabolize imported sugars, the greater is its ability to compete for assimilates being exported by the sources. Such competition determines the distribution of transport sugars among the various sink tissues of the plant (assimilate partitioning), as least in the short term. Of course, events in sources and sinks must be synchronized, so an additional level of control lies in the interaction between areas of supply and demand. Turgor pressure in the sieve elements could be an important means of communication between sources and sinks, acting to coordinates rates of loading and unloading. Chemical messengers are also important in signaling to one organ the status of the other. These chemical messengers include plant growth regulators (hormones) and nutrients, such as sucrose, potassium, and phosphate.

There has been a surge in research on assimilate allocation and partitioning with the goal of improving yields of crop plants. In the past, efforts by plant breeders to increase yield by increasing net photosynthetic rates have generally been unsuccessful. However, significant improvements in yield have resulted from increases in *harvest index*, the ratio of commercial or edible yield to total shoot yield (the latter including inedible portions of the shoot). Clearly, an understanding of partitioning should enable plant breeders and molecular biologists to select and develop varieties with improved transport to edible portions of the plant. As we will see, however, allocation and partitioning in the whole plant must be coordinated, and increased transport to edible tissues must not occur at the expense of other essential processes and structures. Thus increasing the retention of photoassimilates that are normally "lost" by the plant will constitute a necessary, complementary approach to increasing yield. In other words, if losses due to nonessential respiration and leaching or exudation from roots can e reduced, crop yield will also be improved.

Allocation in Source Leaves is Regulated by Key Enzymes

The quantity of sucrose available for export is influenced by a number of biochemical reactions and carrier-mediated events, including the rate of CO_2 fixation. In fact, increasing the rate of photosynthesis in a source leaf generally results in an increase in the rate of translocation from the source. The actual fraction of fixed carbon that is channeled to the so-called transport pool is determined by other processes. Control points include the allocation of triose phosphates to the regeneration of intermediates in the C3 photosynthetic carbon reduction (PCR) cycle, to starch synthesis, or to sucrose synthesis, and the distribution of sucrose between transport and temporary storage pools. A number of

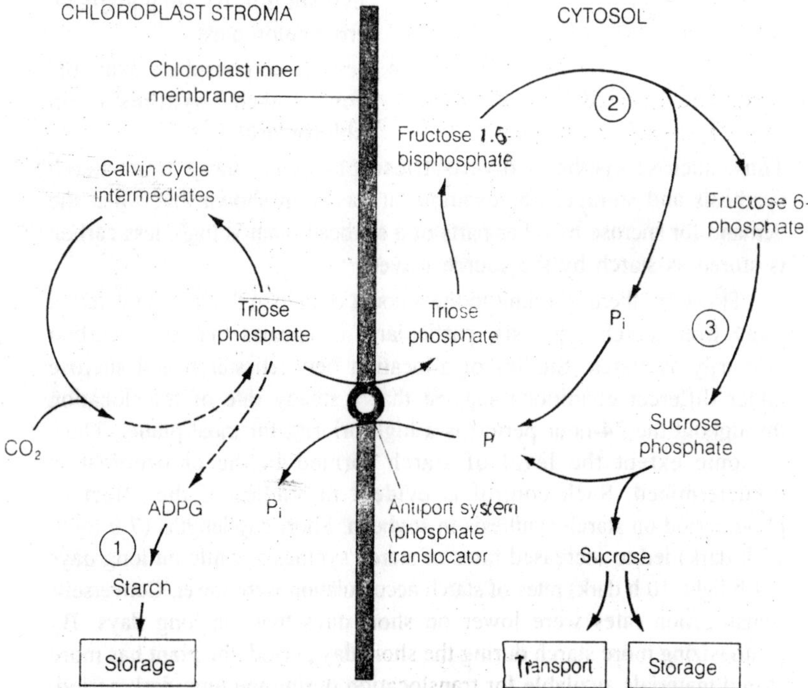

Fig. 5.10. A simplified scheme for starch and sucrose synthesis.

enzymes operate in these pathways, and the controls are quite complex. However, one area of importance is the coordination of starch and sucrose synthesis.

The rate of starch synthesis in the chloroplast must be coordinated with sucrose synthesis in the cytosol. Triose phosphates (glyceraldehyde 3-phosphate and dihydroxyacetone phosphate) produced in the chloroplast by the C3 PCR cycle can be used for either starch or sucrose synthesis (disregarding for the moment utilization of fixed carbon by the photosynthesizing cell itself). Since sucrose synthesis occurs in the cytosol, the triose phosphate destined for sucrose synthesis must leave the chloroplast. This is accomplished by the *phosphate translocator*, located in the chloroplast membrane. This membrane carrier exchanges triose phosphate from the chloroplast for orthophosphate from the cytosol. Whenever sucrose is synthesized, phosphate is released into the cytosol. The entire sequence of events is summarized as follows:

Sucrose synthesis in cytosol → release of phosphate
→ exchange of phosphate from cytosol for triose phosphate from chloroplast
→ less triose phosphate available for starch synthesis in chloroplast

Thus, sucrose synthesis diverts triose phosphate away from starch synthesis and storage. For example, it has been shown that when the demand for sucrose by other parts of a soybean plant is high, less carbon is stored as starch by the source leaves.

However, there is a limit to the amount of carbon that can be diverted away from starch synthesis, particularly in species that store carbon primarily as starch. Studies of allocation between starch and sucrose under different conditions suggest that a steady rate of translocation throughout the 24-hour period is a high priority for most plants. Thus, to some extent the level of starch formed in the chloroplast is predetermined. Such control is evident in studies of the effect of photoperiod on starch synthesis in soybean. Short day lengths (7 h light, 17 h dark) led to increased rates of starch synthesis, while on long days (14 h light, 10 h dark) rates of starch accumulation were lower. Conversely, translocation rates were lower on short days than on long days. By synthesizing more starch during the short-day period, the plant has more stored materials available for translocation during the long dark period. The biochemical basis for such "anticipatory" controls is not known. To some extent, the control of starch synthesis has a genetic basis, as has been observed in studies of isolated cells of various species. Species that tend to accumulate more starch in the light as whole plants retain that capacity even when mesophyll cells are isolated from the remainder of the plant and investigated in cell culture.

Sink Tissues Compete for Available Translocated Assimilate

Translocation to sink tissues depends on the position of the sink in relation to the source and on the vascular connections between source and sink. Another factor determining the pattern of transport in whole plants is competition between sinks. For example, reproductive tissues (seeds) can compete with growing vegetative tissues (young leaves and roots) for available translocate. Competition is indicated by numerous experiments in which removal of a sink tissue from a plant generally results in increased translocation to alternative, presumably competing sinks. The reverse type of experiment involves altering the source supply while leaving sink tissues intact. When the supply of assimilates from sources to competing sinks is suddenly and drastically reduced by shading all the source leaves but one, the sink tissues now depend on a single source. In sugar beet and bean plants the rates of photosynthesis and export from the single remaining source leaf usually do not change over the short term (approximately 8 h; Fondy and Geiger, 1980). However, the roots receives less sugar from the single source, while the young leaves receive relatively more. Presumably, in these plants the young leaves are stronger sinks than the roots. A stronger sink can deplete the sugar content of the sieve elements more readily and thus steepen the pressure gradient, increasing translocation toward itself. An effect on the pressure gradient is also indicated indirectly by experiments in which transport to a sink water potential more negative. Soybean seedlings transplanted to vermiculite containing one-eighth as much water as controls partitioned dry matter from the cotyledons to the roots at a rate twice that of the controls.

Sink Strength is a Function of Sink Size and Sink Activity

Various experiments indicate that the ability of a sink to mobilize assimilates toward itself, the *sink strength*, depends on two factors: *sink size* and *sink activity*.

$$\text{Sink strength} = \text{sink size} \times \text{sink activity}$$

Sink activity is the rate of uptake of assimilates per unit weight of sink tissue, and sink size is the total weight of the sink tissue. Altering either the size or activity of the sink results in changes in transport patterns. For example, less sugar is translocated to an ear of wheat if some of the grains are removed from the ear, lowering its sink size. This reduced translocation is partly due to an inhibition of transport velocity. Changes in sink activity can be more complex, since a number of activities in sink tissues can potentially limit the rate of uptake by the sink. These activities include unloading from the sieve elements, metabolism in the cell wall, retrieval from the apoplast in some cases, and metabolic processes that use the assimilate in either growth or storage. Cooling a sink tissue inhibits activities that require metabolic energy and results in a decreased in the speed of transport toward the sink. Because cooling is so general, however, it does not permit discrimination among processes. Single gene mutations can be helpful here, as in so many other instances. In corn, a mutant that has a defective enzyme for starch synthesis in the kernels transports less material to the kernels than does its normal counterpart. In this mutant, a deficiency in assimilate storage leads to an inhibition of transport. These results on sink size and activity describe sink strength in a very general fashion. Much research remains to be done on the individual steps involved in unloading, so that sites of regulation and control can be identified and the mechanism of competition between sinks can be better understood.

Changes in the Source-to-Sink Ratio bring about Long-Term Alterations in Source Metabolism

When all but one of the source leaves on a soybean plant are shaded for an extended period (e.g. 8 days), many changes occur in the single remaining source leaf, including a decrease in starch concentration and increases in photosynthetic rate, activity of ribulose bisphosphate carboxylase (Rubisco, the CO_2 fixation enzyme of photosynthesis), sucrose concentration, transport from the source, and orthophosphate concentration. Thus, while only the distribution of assimilates among different sinks may change over the short term, over longer periods the metabolism of the source adjusts to the altered conditions. As indicated previously, sucrose and starch synthesis in the source often responds to a change in the sink demand for sucrose. When the sink demand for sucrose is high, sucrose synthesis is increased at the expense of starch synthesis, and vice versa.

Photosynthetic rate (the net amount of carbon fixed per unit leaf area per unit time) often varies over a period of several days in response to altered sink demand, increasing when sink demand increases and decreasing when sink demand decreases. It has been postulated that an accumulation of assimilates in the source leaf could account for the linkage between sink demand and photosynthetic rate. One of the following mechanisms could be operating:

INDEX